The Life of Plants in a Changing Environment

The Life of Plants in a Changing Environment

Edited by

Rishikesh Upadhyay

Cambridge Scholars Publishing

The Life of Plants in a Changing Environment

Edited by Rishikesh Upadhyay

This book first published 2022

Cambridge Scholars Publishing

Lady Stephenson Library, Newcastle upon Tyne, NE6 2PA, UK

British Library Cataloguing in Publication Data
A catalogue record for this book is available from the British Library

ISBN (10): 1-5275-7676-0
ISBN (13): 978-1-5275-7676-6

CONTENTS

List of Illustrations vii

Preface ix

Chapter One 1
The Impact of Various Environmental Factors on Secondary Metabolites in Plants
D Jain, P Chaudhary, R Tripathi and P Janmeda

Chapter Two 34
The Consequences of Oxidative Stress in Plants
G Garg, V Gupta and S Bhati

Chapter Three 62
The Effects of Abiotic Stress on Plants
S. Duhan

Chapter Four 104
The Effects of Heavy Metal Pollution on Plants and their Tolerance Mechanisms
Manju Shri and S. Dubey

Chapter Five 133
Flooding and Submergence Responses in Plants
S. Duhan

Chapter Six 153
The Impact of Plastics on Plant Growth, Agricultural Crops, and Possible Remedies
ND Dhanraj, M Thampi and MS Jisha

Chapter Seven 171
Interactional Effects of UV-B Radiation and Drought Stress on Plants
B Jamra, RS Bapna, S Kataria and M Jain

Chapter Eight 191
Biotic Stress in Plants: A Metabolomics Perspective
S Vijayan, M Thampi and MS Jisha

Chapter Nine 208
Influence of Magnetic Nanoparticles on Plant Growth and Stress Adaptation in Natural Environments
Abhijith P, Bhavya G and Geetha N

Chapter Ten 229
Mobile Radiation Responses in Plants
P. Duhan

Chapter Eleven 239
Programmed Cell Death and Abiotic Stress Responses in Plants
RK Upadhyay and L Darnei

Frequently Used Acronyms 247

Contributors 251

Index 254

List of Illustrations

1. Pathways for the production of variable secondary metabolites in plants 3
2. Different methods for the production of secondary metabolites from plants 5
3. Importance of secondary metabolites to various industries 8
4. Effect of different abiotic stresses in plants 13
5. Effect of light on the production of various secondary metabolites in plants 14
6. Effect of temperature on the production of various secondary metabolites in plants 16
7. Effect of heavy metals on the production of various secondary metabolites in plants 17
8. Effect of drought on the production of various secondary metabolites in plants 18
9. Effect of salinity on the production of various secondary metabolites in plants 20
10. Effect of location on the production of various secondary metabolites in plants 22
11. (a, b) Biological source of reactive oxygen in plants and ROS 36
12. Effects of oxidative stress caused by salinity 42
13. A schematic presentation of the transcriptional regulation of transcription factors playing key roles in cold stress conditions in plants 43
14. Responses of plants to cold stress 44
15. Signal transduction in plants under heavy metal stress conditions 46
16. The solar spectrum encompasses ultraviolet, visible, and infrared radiation 50
17. Ultraviolet-B radiation mediated responses in plants; balancing damage and protection 51
18. Outline of strategies for improving crop performance under pathogenic stress 53
19. Effect of abiotic stresses on plants and their strategies to cope 65
20. Metals acting as cofactors for enzymes involved in various important metabolic reactions 75
21. Responses of plants to flooding and submergence stress 135

22. Aerenchyma formation in 3 genotypes of 20-day-old pigeon pea plants (ICPH 2431, UPAS 120, and PARAS) 1 day after removal from an 8-day waterlogging treatment 140
23. Hypothesized effects of different types of microplastic particles on plants and potential food safety concerns for crops 154
24. Polymer materials used in everyday life 156
25. Overview of plastics used in agriculture.. 157
26. Plastics found in natural substances... 161
27. Concept diagram showing the various mechanisms via which microplastic particles can affect plant growth 165
28. Responses of plants to various abiotic stresses 174
29. Antioxidative system in response to UV-B and drought stress......... 180
30. The plant metabolomics mechanism for crop improvement............. 194
31. Genome-wide association studies (GWASs) assisted by metabolomics techniques (mGWAS) and metabolic quantitative trait loci (mQTLs) are powerful tools to detect genetic variations linked with metabolic traits in plants... 199
32. Alignment of magnetic moments in different materials under an applied external magnetic field 213
33. Scanning electron micrograph of metallic nanoparticles at 15000 x magnification 215
34. Coating and functionalization of magnetic nanoparticles 217
35. Helmholtz coil system 219
36. The role of MNPs in plant growth and stress management 222
37. Radiation emitted by cellphones.. 231
38. Mobile phone radiation - effects on plants....................................... 232
39. Abiotic stress responses to programmed cell death in plants............ 242
40. Outline of programmed cell death in plants..................................... 244

PREFACE

Today, plants are faced with many new challenges. The life and growth of plants is in decline due to continuing changes to the environment, mostly caused by different types of stress effects, an increasing human population, and changing food habits. Changes to habitats, stress adaptations, and climate change create new problems with respect to the morphological, physiological, and biochemical aspects of plants and their ability to adapt to these changes during their natural life cycles.

There is no doubt that improvements in the genetic and molecular biology of plants (including crop species) will be key to increasing growth, production, and yields, or changing plant adaptations in response to stress in the future.

The dynamic and increasing body of knowledge concerning the effects of changing environments, stress factors, and their impact on plants and crops has resulted in the compilation of this unique and comprehensive collection, which deals with plants' changing situations and the stresses imposed on them (including vegetable species), in the form of this book for advanced undergraduates, research scholars, teachers, and professionals in the fields of plant adaptation, ecology, and environmental science.

This book consists of 11 chapters prepared by 25 authors from around the country. Chapter 6 emphasizes the impact of plastics on plants and crop productivity, with the hope that this will result in better understanding of plant adaptations to the changing environment.

The induction of oxidative stress in plants is an integral part of plant responses to different environments. Chapter 2 provides knowledge concerning plant growth in relation to oxidative stress in our changing environment. Chapters 4 and 5 focus on how plants adapt and tolerate heavy metal pollution, and how flooding and submergence tolerance is regulated under natural environments during plant life cycles.

Chapters 1 and 7 address different environmental factors affecting secondary metabolites and the interactional effects of UV-radiations and drought.

Chapter 3 emphasizes different aspects of the abiotic stress response. Chapter 8 tackles the intricacies of plant responses to biotic stress with special reference to the metabolomics approach. Chapters 9 and 10 give an overview of how plants react to magnetic nanoparticles and mobile phone radiation in natural environments. Finally, Chapter 11 covers programmed cell death and plant responses to abiotic stress conditions.

Numerous figures and tables appear in this book to facilitate comprehension of the presented material. This book also includes a comprehensive index, a list of illustrations, and a list of acronyms used to further increase the accessibility of the information presented.

All the chapters have been written by experts with extensive experience in their fields of expertise. I am indebted to all the authors for their excellent contributions, which make this book, I think, a valuable resource for different aspects of plant responses to our changing environment. I am aware of the fact that not all the relevant topics in this area could be included in this book due to size limitations.

Finally, I sincerely appreciate the invaluable efforts of each of the contributors who responded to my request for contributions to this volume. Their proficiency and knowledge in their areas of expertise has made this important task possible.

I also wish to acknowledge Rebecca Gladders and James Brittain, who coordinated the publishing process, and Helen Edwards, who has been particularly supportive and helped at various stages of the preparation of this manuscript for publication. I am deeply indebted to all of these people.

Thank you.

RISHIKESH UPADHYAY,
FSASS, PhD, MPhil.
Haflong, Assam, INDIA

CHAPTER ONE

THE IMPACT OF VARIOUS ENVIRONMENTAL FACTORS ON SECONDARY METABOLITES IN PLANTS

D JAIN, P CHAUDHARY, R TRIPATHI AND P JANMEDA

Climate change is responsible for causing changes in various bioactive constituents of plants, including medicinal plants and food crops, across the globe. A number of studies have determined that climate change is not only affecting crop yields, but also bringing about variation in the concentration of secondary metabolites. Plant secondary metabolites (PSM) involve different classes of naturally available compounds and variable biochemical pathways synthesize them under the influence of environmental factors and herbaceous predators. With the help of secondary metabolites, plants can adapt to their changing environments and stress conditions. Despite their significant biological activity, these secondary metabolites are broadly employed in a wide range of industries—as food additives, pesticides, fragrance compounds, cosmetics, and therapeutics. Environmental factors, such as light intensity, humidity, temperature, salinity, and location, etc., are all very important. These factors can be influential through different mechanisms that produce variations in the biogenesis and accumulation of secondary metabolites. Alterations to any of these factors can bring about changes in the content of secondary metabolites. A good understanding of the mechanisms involved in the accumulation, degradation, and synthesis of metabolites is needed for the formulation of future strategies to enhance the productivity, safety, and reliability of secondary metabolites in plants. In this chapter, we present a detailed overview of the possible role of environmental factors in the instability of secondary metabolites in plants.

Introduction

Secondary plant metabolites can be described as compounds that play no specific role in the regulation of life processes in the host plant; they are, however, required by the plant for interacting with external stimuli in defense and adaptation (Isah et al., 2019). In crop plants, different types of secondary metabolites (SM) are devised from primary metabolites with different physiological functions.

These SMs play an important role in the establishment of a strong relationship between plants and the environment in terms of their fitness and survival, making these compounds valuable as primary metabolites (Kliebenstein et al., 2012). They are used by the host plant to provide a defense against pathogenic organisms and herbivores (War et al., 2012). They also contribute to the dynamic colour, taste, and odour of plants. Secondary metabolites are a common source of material for pharmaceuticals, flavourings, food additives, and other related industrial uses. Chemicals such as nitric acid, jasmonates, polyamines, salicylic acid, abscisic acid, and calcium are used in the stress responses of the host plant. The generation of these compounds is limited and mainly reliant on the developmental and physiological stage of the host plant (Ramakrishna et al., 2011).

Abiotic stresses influence plant metabolism. When crop plants are challenged by abiotic stresses, the results include a reduction in various morphological features, such as root volume, number of branches, leaf area, leaf number, and height, etc., further affecting the biomass of the plants (Fahad et al., 2017). In normal environmental conditions, the production of bioactive metabolites is low, whereas metabolite production is greater when plants encounter abiotic stresses. The accumulation of phenolic and terpenoid compounds in plants has been found to increase under stress conditions. The concentration of various other metabolites is strongly dependent on the growing conditions. By manipulating the metabolic pathways and expression of certain genes associated with the expression of natural bioactive compounds, the synthesis of secondary metabolites has been shown to be heightened in response to different abiotic stresses (Kumar et al., 2018). As such, this chapter aims to summarize the biosynthesis, creation, production, and importance of variable secondary metabolites in host plants in response to different abiotic stresses.

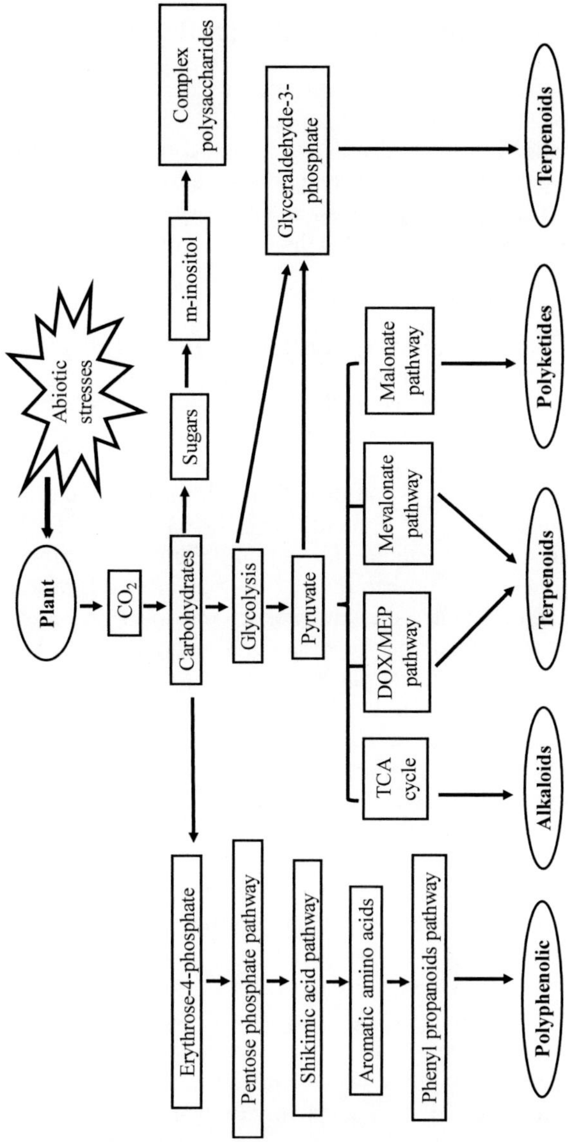

1. Pathways for the production of variable secondary metabolites in plants.

Expression of secondary metabolites in plants

The secondary metabolites of plants can be categorized into three types on the basis of their synthesis: terpenoids, phenolics, and alkaloids. It has been determined that terpenoids are synthesized via the mevalonic and DOX/MEP pathway, phenolic compounds are formed via the malonic acid and shikimic acid pathway, and alkaloids are synthesized from an aromatic amino acid in the shikimic acid pathway and from aliphatic amino acids in the tricarboxylic acid cycle (Parsaeimehr et al., 2011) [**1**].

Creation of secondary metabolites in plants

A number of components, including the living creatures present, the edaphic conditions, and the season/atmosphere, have been found to affect the presence or lack of some secondary metabolites in medicinal plants (Sampaio et al., 2016). Another factor influencing the formation of secondary metabolites is the interaction of insects and plants. For example, many medicinal plants require the help of pollinators to cross-pollinate. In an open space, the wind can do the job, but in woods where herbs and bushes are found under the canopies of the trees, there is insufficient breeze for their fertilization. To attract pollinating bugs, the plant produces a pleasant smell and provides dust and nectar as food. Dust and pollens are a common source of nourishment for birds and insects (Brittain et al., 2014) and many plants contain an advanced blossom part for fertilization with the help of bugs. In return, these creatures reside on the plants. They lay their eggs on them and when they hatch, the hatchlings start feeding on the leaves. The activity of these hatchlings disturbs the plant and results in the production of some dangerous compounds (Jabr et al., 2013). As such, unfavourable abiotic conditions stress the plants and in return the plants respond to these stresses by producing certain secondary phytochemicals (Biology et al., 1988).

Production of secondary metabolites

The secondary metabolites of a plant are a unique resource for useful chemicals, food additives, and pharmaceuticals. They are employed in other areas also. Direct extraction and chemical synthesis are commonly used to get these metabolites. Plant cell cultures and metabolic engineering have been suggested as potential alternatives for the production of metabolites that are hard to obtain from extraction and synthesis (Tiwari et al., 2015) [**2**].

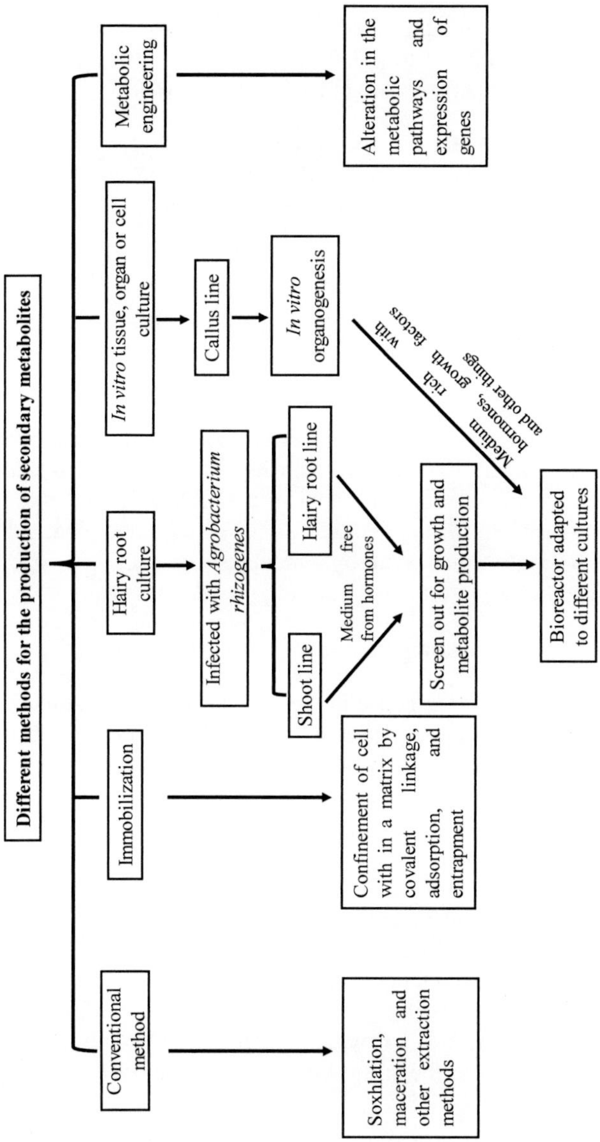

2. Different methods for the production of secondary metabolites from plants.

- **Conventional method**

  The conventional method is based on the extraction of phytochemicals from plant tissues using various procedures such as supercritical and solvent steam extraction.

The latest advancements in fermentation technology, enzymology, and molecular biology have shown that these systems are important for the *in vitro* production and synthesis of secondary metabolites (Hussain et al., 2012). The main process includes:

- **Immobilization method**

  Confinement of biocatalyst or cell within a suitable matrix by covalent linkage, adsorption, and entrapment. Under favourable physicochemical conditions, and with the addition of a specific substrate, the desired bioactive constituents are synthesized. Immobilization with an appropriate bioreactor system allows many advantages, such as the continuous operation of the process, but for the creation of an immobilized plant cell culture, artificially or naturally induced seepage of accumulated metabolites into the neighboring medium is required.

- **In vitro tissue, organ, and cell culture methods**

  Tissue and plant cell cultures can be routinely developed under aseptic conditions from explants, from samples of meristem, roots, stem, and leaves, etc., both for the extraction and the multiplication of secondary metabolites. Hairy root, cell suspension, callus, root, and shoot cultures can all be utilized for the creation of metabolites of interest. Localized metabolites can be produced through suspension or callus cultures, whereas for those metabolites confined to a specific gland or part, organ cultures and differentiated micro-plants can be chosen. The quantity of secondary metabolite formation in cell cultures can be improved by treating the plant cells with abiotic and biotic elicitors. Yeast extract, fungal carbohydrate, and methyl jasmonate are the primarily utilized elicitors (Thirumurugan et al., 2018).

- **Hairy root cultures**

  Hairy root cultures involve the inoculation of plant roots with *Agrobacterium rhizogenes* and other microorganisms (arbuscular mycorrhizal fungi) for the production of secondary metabolites. The phenotype of the hairy root is characterized by genetic stability, lateral branching, a lack of geotropism, fast growth, and hormone-independent synthesis. The secondary metabolites obtained after the infection of the host plant with *A. rhizogenes* and AMF are similar to those that are synthesized in the intact roots of the plant and with the same or even greater yield (Hussain et al., 2012; Chaudhary et al., 2019).

- **Metabolic engineering**

  Metabolic engineering involves making purposeful and targeted changes to metabolic pathways for the better understanding and use of these pathways for supramolecular assembly, energy transduction, and chemical transformation. In a number of cases, the number of secondary metabolites is too low to be used for commercial purposes and metabolic engineering provides several strategies to decrease catabolism, block competitive pathways and feedback inhibition, increase the flux of carbon through the overexpression of certain genes in biosynthetic pathways, and upgrade the productivity of metabolites in plants (Hussain et al., 2012; Gonçalves et al., 2018).

Importance of secondary metabolites

Plants struggle to survive for an extended period during their establishment. Gradually they gain the ability to resist stressful conditions through the production of different types of secondary metabolite with variable bioactivities. These metabolites play a crucial role in the pharmaceutical, food, and dye industries, defend the host plant against various pathogenic organisms with other competing plants, and facilitate reproduction and pollination. Some essential secondary metabolites are listed in [**3**].

3. Importance of secondary metabolites to various industries.

Plant species	Secondary metabolites	Uses/Applications	References
Cactus and *Linum usitatissimum*	Mucilage	Demulcent	Hussein *et al.*, 2018
Althaea officinalis	Marshmallow	Cough suppressant	
Origanum compactum, *Coriandrum sativum*, *Artemisia herba-alba*, *Cinnamomum camphora*, *Mentha piperita*	Carvacrol and thymol, Linalol, α and β-thuyone and camphor, 1,8-cineole, Menthol and menthone	Used as fragrances in food industry, local anesthetic remedy, spasmolytic, anti-inflammatory, sedative, analgesic, antimicrobial and antiseptic	
Simmondsia chinensis	Liquid wax and jojoba wax	Wound healing, anti-aging, and anti-inflammatory activity	
Linum usitatissimum	Fixed oil	Decreases the risk of cardiovascular diseases and atherosclerosis	
Quassia amara	Quassinoids	Insecticidal property	
Boswellia carterii	β-boswellic acid and α-boswellic acid	Anti-rheumatic and anti-inflammatory activity	
Glycyrrhizin glabra	Glycyrrhizin	Antitussive agent, treat cirrhosis, cheonic and hepatitis	
Bupleurum falcatum	Saponins	Anti-inflammatory property	
Catharanthus roseus G.	Vinblastine	Diabetes	

Catharanthus roseus L.	Vincristine	Nephroblastoma, lymphoma, neuroblastoma, rhabdomyosarcoma, and acute lymphatic leukemia	Seca *et al.*, 2018
Vicia ervilla, Citrus aurantium, Glycyrrhiza glabra	Apigenin	Anti-infection, antiviral, and anticarcinogenic activity	Parsaeimehr *et al.*, 2011
Fragaria spp.	Fisetin	Anti-oxidant, anti-carcinogenic and anti-inflammatory activity	Parsaeimehr e*t al.*, 2011
Taxus brevifolia Nutt.	Paclitaxel	Anticancer activity in the cure of lung and breast cancer	Seca *et al.*, 2018
Euphorbia peplus L.	Ingenol mebutate	Antitumor component	Seca *et al.*, 2018
Curcuma longa L.	Curcumin	Chemo sensitizing, chemotherapeutic, anti-oxidant, and anti-inflammatory activity	Seca *et al.*, 2018
Chrysanthemum species	Terpenes	Commercial insecticides	Pagare *et al.*, 2015
Rauwolfia serpentine	Reserpine	Treatment of thyrotoxicosis, tachycardia, and hypertension	Lobay *et al.*, 2015
Cinchona tree	Quinine	Anti-malarial	Achan *et al.*, 2011
Cassia tora L.	Rotenoids	Larvicidal activity	Vats *et al.*, 2018
Arctostaphylos uvaursi	Phenolic content	Diuretic and antimicrobial activity	Hussein *et al.*, 2018

Capsicum spp.	Capsaicinoids	Analgesic, rubefacient, and circulatory activity	Hussein *et al*., 2018
Vaccinium oxycoccos	Tannin-containing Juice	Urinary antiseptic	Hussein *et al*., 2018
Atropa belladonna, Datura stramonium, Daphne mezereum, Ruta graveolens, Aesculus hippocastanum	Coumarins	Anti-alzherimer's, anti-cancer, anti-coagulant, and anti-inflammatory activity	Hussein *et al*., 2018
Polygala nyikensis	Xanthones	Antifungal activity	Hussein *et al*., 2018
Cedrus deodara	Wikstromal, matairesinol, dibenzyl butyrolactol	Cytotoxic activity	Hussein *et al*., 2018
Solanaceae plant family	Nicotine	Anti-inflammatory, insecticidal, antiherbivore, and stimulant activity	Kabera *et al*., 2014
Tomato plants	Tomatine	Antifungal, anticancer and immune effect	Kabera *et al*., 2014
Papaver somniferum	Codeine	Act as CNS, used in acute pulmonary edema	Kabera *et al*., 2014
Polypodium leucotomos Hook.	Caffeic acid, ferulic acid, and chlorogenic acid	Used as sunscreen	González-Minero *et al*., 2018
Camellia sinensis L.	Vitamin C	Inhibit peroxidation of lipid	González-Minero *et al*., 2018

Vitis vinfera L.	Polyphenols	Inhibit peroxidation of lipid	González-Minero *et al.*, 2018
Talaromyces verruculosus	Red pigment	Textile industry	Chadni *et al.*, 2017
Vanilla planifolia	Vanillin	Flavoring agent	Kallscheuer *et al.*, 2018
Curcuma longa	Curcumin	Flavoring agent	Kallscheuer *et al.*, 2018

Influence of environmental stress on plant development

Across the globe, plants are subject to different abiotic stresses, which affect agricultural productivity. These abiotic factors are associated with each other and can appear in the form of plant cell homeostasis, malnutrition due to ion distribution, and osmotic stress. Productivity and growth rates are greatly impacted due to alterations in the expression of groups of genes. As such, recognizing which genes are responsible for the control of abiotic stress is necessary to understand the influence of these abiotic stresses on crop plants (Gull et al., 2019).

Environmental factors

Plants are affected by various environmental factors in achieving the sustenance necessary for life and are thus affected by different abiotic and biotic stimuli, leading to the production of particular bioactive constituents (Zhi-lin et al., 2007). Plants of similar species that are cultivated in variable environmental conditions may show variation in the yield of secondary metabolites. Biotic and abiotic factors cause biotic and abiotic stresses, respectively, in host plants as a result of unfavourable conditions. In order to counteract these stresses, plants produce dynamic secondary metabolites. As a result, environmental factors are key factors driving the synthesis of plant metabolites. Abiotic stresses are caused by the composition and type of soil, temperature, water availability, and light intensity, etc., which all influence the productivity and quality of plants.

In contrast, biotic stress is induced by living organisms such as parasites, fungi, viruses, and bacteria (Radusiene et al., 2012). Chemicals, such as growth regulators (NAA, IAA, and 2,4-D), metals (Mn, Zn, Fe, Cr, Co, Cd, and Ni, etc.), pesticides, pollutants, and gaseous compounds and radiation (UV and simple light) also contribute to abiotic stress in host plants (Ramakrishna et al., 2011). Plants do not acquire an immune system and are not able to move like animals. They gain tolerance of these unfavourable conditions through the accumulation of bioactive metabolites.

Impact of environmental stress factors in plants

Abiotic stresses constitute the primary limiting factor for the sustainable productivity of crops. Plants overcome these adverse effects, born of the edaphic and environmental conditions, through intrinsic biological mechanisms. Plants require minerals, nutrients, carbon, water, and light for reproduction, development, and growth. Adverse conditions (above or below the optimal level) restrict plant development and growth. Unfavourable factors in an ecosystem include salinity, drought, and low or high temperatures, all of which create complex stress conditions. Plants respond to these stresses in a variety of ways (Ahmad et al., 2015; Jiang et al., 2016). Commonly, the effects of stressful growing conditions appear at the cellular level first, before appearing at the physiological level. Prolonged stress due to the unavailability of water decreases leaf size, stomatal openings, and water potential; delays fruiting and flowering; restricts plant productivity and growth; suppresses root growth; and reduces the viability, number, and size of seeds (Xu et al., 2016). Exposure to high and low light intensity reduces physiological processes and unfavourably influences the development and growth of plants. Extreme light intensity increases the production of reactive oxygen species, which manipulate enzymes and other biomolecules (Li et al., 2009). Significant increases and decreases in temperature are primary causes of loss of productivity (Pareek et al., 2010). Several edaphic factors, such as anthropogenic perturbation, contamination by pollutants, and alkalinity, acidity, and salinity of soil greatly affect the development and production of crops (Bui et al., 2013; Emamverdian et al., 2015) [**4**].

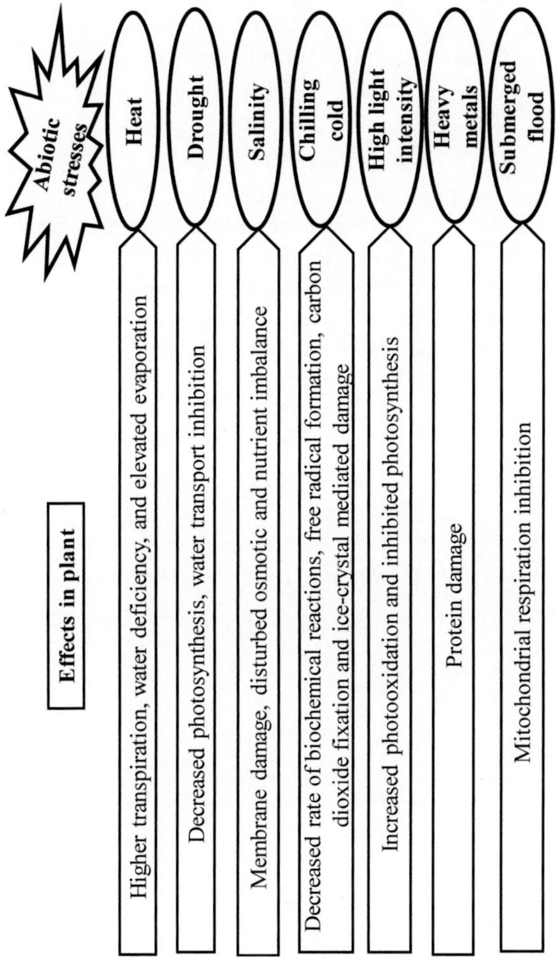

4. Effect of different abiotic stresses in plants.

Abiotic factors

Throughout their ontogeny, plants connect with their environment and come into conflict with numerous components, including light, soil, water, temperature, and chemical substances like fertilizers and minerals, which are all necessary for plants to grow and survive properly. However, plants are also impacted by a greater or lesser number of these abiotic factors, contributing to differences in their development or accumulation of PSMs.

1. **Light**

This factor concerns photoperiod (duration), direction (frequency or wavelength), and intensity (quantity). Reacting to light exposure, plants receive significant economic and consumption benefits from various secondary metabolites, including flavonoids, triterpenoids, and phenolic compounds due to their well-recognized antioxidant properties. A short daytime of light exposure will result in a reduction of about 40 % of caffeoylquinic acids and also a roughly double reduction in the content of flavonoid aglycones compared to a long daytime of light exposure (Yang et al., 2018). As such, light intensity has a substantial influence on the accumulation of PSMs, as listed in [**5**].

5. Effect of light on the production of various secondary metabolites in plants.

Plant species	Compounds	Conditions	Reference
M. glomerata	Coumarins	Light irradiation	de Castro *et al*., 2006
Artemisia annua	Artemisin	Light irradiation	Liu *et al*., 2002
Panax quinquefolius	Higher levels of ginsenosides	Longer light exposure	Fournier *et al*., 2003
Vaccinium myrtillus	Flavonoid biosynthesis pathway activation	Continuous solar radiation	Jaakola *et al*., 2004
Catharanthus roseus	Vinblastine, vincristine, Flavonoid	UV-B light	Bernard *et al*., 2009; Dixon and paiva, 1995
Vitis vinifera	Stilbene	UV-C irradiation	Wang *et al*., 2010; liu *et al*., 2010
Centaurea cyanus	Anthocyanins	UV-C irradiation	Kakegawa *et al*., 1991
Hordeum vulgare	Flavonoids	UV-B	Liu *et al*., 1995

Cucumis sativus	Polyamines	UV-B	Kramer *et al.*, 1991
Picea abies	Flavonols	UV-B irradiation	Fischbach *et al.*, 1999
Digitalis purpurea	Digitoxin	Light irradiation	Hagimori *et al.*, 1982
Melastoma malabathric	Anthocyanins	Light irradiation	Chan *et al.*, 2010
Ocimun basilicum	Rosmarinic acid accumulation	Exposure to red light (600–700 nm)	Shiga *et al.*, 2009
Zingiber officinale	Gingerol and zingiberene	Light	Anasori *et al.*, 2008
Taxus cuspidate	Taxol and baccatin III	White light	Fett-Neto *et al.*, 1995

2. Temperature

Good plant development relies on an ideal environment. Lower and higher temperatures have a detrimental influence on plants commsurate with heat and cold tension (Yadav, 2010). Several studies have shown an improvement in PSM biosynthesis as well as a decline in secondary metabolite in reaction to high temperatures. Cold stress concerns low temperatures (< 20℃), which impact plant growth and development and considerably limit productivity (Chinnusamy et al., 2007). Low temperatures retard plant growth, contributing to a decreased rate of photosynthesis and a lower yield of *Capsicum annuum*. Low temperatures often contribute to specific molecular, physiological, and biochemical alterations in plants that render them immune to cold exposure, which is known as cold acclimation (capacity to thrive and avoid injury under low-temperature stress). Research has revealed that the pressure of cold has a significant impact on variation in PSM content, as listed in [**6**].

6. Effect of temperature on the production of various secondary metabolites in plants.

Plant species	Compounds	Conditions	Reference
Panax quinquefolius, P. ginseng	Ginsenosides	High temperature	Yu *et al.*, 2005; Jochum *et al.*, 2007
Hypericum perforatum	Hypericin, hyperforin	Low temperature	Zobayed *et al.*, 2005
Hemerocallis sp.	Suberin or lignin	-	Griffith *et al.*, 2004
Pinus pinaster	Endogenous jasmonates	-	Pedranzani *et al.*, 2003
Rhodiola crenulata	Melatonin	-	Zhao *et al.*, 2011
Medicago sativa	Putrescine	-	Nadeau *et al.*, 1987
Melastoma malabathricum	Anthocyanin	-	Chan *et al.*, 2010
Artemisia spp.	Higher levels of artemisin	Lower temperature	Wallaart *et al.*, 2000; Brown *et al.*, 2010
Nicotiana tabacum	Higher levels of anthocyanins	Lower temperature	Huang *et al.*, 2012

3. Heavy metals

Heavy metals can induce shifts in the metabolism of plants and affect the production of photosynthetic pigments, proteins, sugars, and non-protein thiols. Metals can enhance the development of bioactive compounds by modifying the dimensions of secondary metabolism (Verpoorte et al., 2002). Several metals, like iron (Fe), cobalt (Co), silver (Ag), and nickel (Ni), have been shown to exhibit the synthesis of secondary metabolites in plant varieties (Zhao et al., 2001). The production of secondary metabolites is affected by various metal ions, such as oxalate, Eu^{3+}, La^{3+}, Cd^{2+}, and Ag^{+}, etc. (Marschner, 1995), as listed in [**7**].

7. Effect of heavy metals on the production of various secondary metabolites in plants.

Plant species	Compounds	Reference
Taxus chinensis	Taxol	Groppa *et al.*, 2001
Amaranthus caudatus	β-cyanins	Obrenovic *et al.*, 1990
Lepidium sativum	Lepidine	Saba *et al.*, 2000
Perovskia abrotanoides	Tanshinone	Arehzoo *et al.*, 2015
Datura stramonium	Sesquiterpenoid, lubimin	Threlfal *et al.*, 1988; Furze *et al.*, 1991
Brassica juncea	35 % increase in oil content	Singh *et al.*, 2005
Lithospermum sp.	Shikonin	Mizukami *et al.*, 1977
Digitalis lanata	Digitalin	Ohlsson *et al.*, 1989
Beta vulgaris	Betalains	Trejo-Tapia *et al.*, 2001
Brugmansia candida	Scopolamine, Hyoscyamine	Angelova *et al.*, 2006
Salvia castanea	Tanshinone	Li *et al.*, 2016
Datura metel	Atropine	Shakeran *et al.*, 2015
Vitis vinifera	Resveratrol	Cai *et al.*, 2013
Ammi majus	Xanthotoxin	Purohit *et al.*, 1995
Bacopa monnieri	Bacoside	Sharma *et al.*, 2015
Beta vulgaris	Betalain	Savitha *et al.*, 2006
Dioscorea bulbifera	Diosgenin	Narula *et al.*, 2005
Atropa belladonna	Tropane alkaloids	Lee *et al.*, 1998
Hyoscyamus albus	Phytoalexin	Mader *et al.*, 1999

4. Water

Water is an essential molecule of plant physiology, serving as the means of transportation of active ingredients and nutrients. As water flow is reduced, or transpiration in the plant rises, water stress is induced, i.e. from drought and salinity stress.

4.1 Drought

Drought sees a lack of water taken up by a plant with a reduction in water capability and turgidity in such a way that its physiological and pathological activities are affected (Tippmann et al., 2006; Lisar et al., 2012). It inhibits profitability, biosynthesis, and photosynthesis, changing the plant's biochemical characteristics (Zobayed et al., 2007; Aimar et al., 2011). Several PSMs encourage plants to flourish. The growth of PSM production in several medicinal plants, such as *Hypericum perforatum*, *C. roseus*, and *Artemisia annua*, has been observed as a consequence of extreme drought stress, as listed in [**8**]. Studies have shown that water plays a critical role in plant metabolism and physiological activities and can modify the biosynthesis and concentrations of PSMs (Azhar et al., 2011).

8. Effect of drought on the production of various secondary metabolites in plants.

Plant species	Compounds	Conditions	References
Trachyspermum ammi	Increases chlorophyll and total phenolic content	Water scarcity	Azhar *et al.*, 2011
Matricaria chamomilla	Reduce oil content	Salinity and drought stress	Razmjoo *et al.*, 2008
Artemisia	Artemisinin	Scarcity of water increases	Zobayed *et al.*, 2007
Hypericum brasiliense	Betulinic acid, quercetin, and rutin	Scarcity of water increases	Zobayed *et al.*, 2007
G. longituba	Total flavonoids content	Water deficiency	Zhang *et al.*, 2012
Ocimum basilicum and Ocimum americanum	Carbohydrates, essential oil, proline, nitrogen, Phosphorus, potassium and protein content	Water stress	Khalid *et al.*, 2006
Salix sp.	Flavonoids, phenolics	Drought	Larson *et al.*, 1988
Stevia rebaudiana	Steviol glycosides	Polyethylene glycol induce drought	Pratibha *et al.*, 2015
Hypericum adenotrichum	Hypericin, pseudohypericin	Polyethylene glycol induce drought	Omer *et al.*, 2013

Hypericum perforatum	Hypericin, hyperforin	Water and osmotic imbalance	Pavlik *et al.*, 2007
Bupleurum Chinese	Saikosaponins	Water stress	Zhu *et al.*, 2009
Salvia miltiorrhiza	Salvianolic acid	Water stress	Liu *et al.*, 2011
Prunella vulgaris	Rosmarinic, ursolic, and oleanol	Drought stress	Chem *et al.*, 2011
Glycyrrhiza uralensis	Glycyrrhizic acid	Water stress	Li *et al.*, 2011
Glycine max	Trigonelline	Drought stress	Cho *et al.*, 2003
Brassica napus	Glucosinolates	Drought stress	Jensen *et al.*, 1996
Lupines angustifolius	Chinolizidin alkaloids	Drought stress	Christiansen *et al.*, 1997
Camellia sinensis	Epicatechins	Drought stress	Hernaendez *et al.*, 2006
Hypericum brasiliense	Betulinic acid, rutine	Drought stress	De Abreu *et al.*, 2005
Helianthus annuum	Chlorogenic acid	Drought stress	De moral *et al.*, 1972
Salvia miltiorrhiza	Rosmarinic acid	Drought stress	Liu *et al.*, 2011
Papaver somniferum	Morphine alkaloids	Drought stress	Szabo *et al.*, 2003
Quercus ilex	Lower monoterpene emissions	Drought stress	Lavoir *et al.*, 2009

4.2 Salinity

Increased soil salinity and many other factors, like heavy rainfall, and low soil temperature create stress for plants, despite water accessibility. High salt concentration, especially due to ions such as Na^+ in the soil, can lessen water and nutrient absorption, as well as the growth, production, photosynthesis, and respiration of plants. This can dehydrate the plant's cell membranes due to the removal of water from the cytoplasm, which causes oxidative pressure (Tippmann et al., 2006). The secondary metabolite concentrations in plants may be attributed to variations in ionic and osmotic stress caused by salt tolerance (Ramakrishna et al., 2011). Many plants, such as *M. pulegium*, *Mentha suaveolens*, *Origanum vulgare*, *Mentha piperita*, *Majorana hortensis*, *Thymus maroccanus*, *M. chamomilla*, *Salvia officinalis*, and *T. ammi* show a reduction in essential oil production under salt stress. In contrast, some plants, like *Satureja hortensis*, *S. officinalis*, and *Matricaria recutita* show an increase in the production of essential oil content under salt stress (Said-Al Ahl et al., 2011). These studies have clearly shown that salinity encourages PSM accumulation, as listed in [**9**].

9. Effect of salinity on the production of various secondary metabolites in plants.

Plant species	Compounds	Conditions	Reference
Rauvolfia tetraphylla	Reserpine	Salt stress	Said-Al Ahl *et al.*, 2011
C. roseus	Vincristine alkaloids and indole alkaloids		Said-Al Ahl *et al.*, 2011; Misra *et al.*, 2006; Fatima *et al.*, 2015
Ricinus comunis	Ricinine alkaloids		Said-Al Ahl *et al.*, 2011
Solanum nigrum	Solasodine		Said-Al Ahl *et al.*, 2011
M. chamomilla	Phenolic acids (protocatechuic, chlorogenic and caffeic acids)		Kovácik *et al.*, 2009
Nigella sativa	Enhancement of Phenols		Bourgou *et al.*, 2010
Menthe pulegium	Enhancement of Phenols		Queslati *et al.*, 2010
Plantago ovata	Proline, flavonoids and saponins		Haghighi *et al.*, 2012

Lycopersicon esculentum	Sorbitol, jasmonic acid		Tari *et al.*, 2010
Datura innoxia	Tropane alkaloids		Brachet *et al.*, 1986
Trifolium repens	Glycine betaine		Varshney *et al.*, 1988
Datura innoxia	Total alkaloid		Brachet *et al.*, 1986
Hordeum vulgare	Flavonoids		Ali *et al.*, 2003
Grevillea ilicifolia, G. arenaria	Anthocyanin		Kennedy et al., 1999
Triticum aestivum	Glycine betaine		Krishnamurthy *et al.*, 1989
Sesamum indicum	GABA		Bor *et al.*, 2009
Cakile maritime	Polyphenol		Ksouri *et al.*, 2007
Glycine max	Trigonelline		Cho *et al.*, 1999
Oryza sativa	Polyamine		Krishnamurthy *et al.*, 1989

5. Location

The location may change the composition and concentration of secondary metabolites in plant species and therefore obstruct particular functions, activities, and behaviours of plants (Badoni et al., 2009; Yang et al., 2004). The impact of location has been observed on the antibacterial activity of *Mentha spicata* grown in diverse areas and showing different outputs (Ullah et al., 2012). Differing secondary metabolite concentrations in the same plant species may depend on the availability of favourable or unfavourable atmospheric changes in different areas, as listed in [**10**]. These studies suggest, specifically, that the locational and external conditions positively influence the biosynthesis of PSMs.

10. Effect of location on the production of various secondary metabolites in plants.

Plant species	Compounds	Conditions	Reference
Nothapody tesnimmoniana	Camptothecinn (monoterpene indole alkaloid)	Climatic conditions	Namdeo *et al.*, 2010
C. roseus	Phenolic compounds	Climatic conditions	Kumar *et al.*, 2013
Matricaria chamomilla	Higher flavonoids, and phenolic acids	Higher altitudes	Ganzera *et al.*, 2008
Leontodon autumnalis	Shifting flavonoids contents	Altitude	Grass *et al.*, 2006, Zidorn *et al.*, 2001

Conclusions and future perspectives

This chapter has shown how abiotic stress factors direct the growth and bioactive metabolite production of host plants. The effects are notable. Sustainable productivity is also affected due to alterations in the ecosystem, for example, the impact of climate transformation on soil microflora, butterflies, bees, etc., also influence plant productivity, adaptation, and ontogeny. However, comprehensive studies are lacking. Above all, climate alteration has been seen to adversely affect soil conditions, salinity, and water availability, with direct impact on the yields of natural products. The main benefit of cell cultures and metabolic engineering is that they involve the synthesis of metabolites, independently of soil and climatic conditions. The utilization of *in vitro* cell culture techniques for the production of pharmaceutical and chemical products has made significant progress. Hence, information gathered through isolated studies needs to be translated into extensive investigation of the influence of climate change. Remedial actions are difficult to recommend, but are necessary alongside preventive approaches at the global scale, since abiotic changes do not have geographical and political boundaries. As such, the utilization of metabolic and biochemical markers needs to be undertaken to determine the impact of abiotic stresses on SM production and plant biomass. Furthermore, the exploration of physiological, metabolic pathways responses to different stresses could require the employment of the metabolic engineering approach to improve the generation of secondary metabolites through the application of next-generation approaches.

References

Achan, J, Talisuna, A, Erhart, A, Yeka, A, Tibenderana, J, Baliraine, F, Rosenthal, P, and Alessandro, U. 2011. "Quinine, an old anti-malarial drug in a world: role in the treatment of malaria". *Malaria Journal*, No. 10: 144.

Ahmad, P, Hashem, A, Abd-Allah, EF, Alqarawi, AA, John, R, Egamberdieva, D, and Gucel, S. 2015. "Role of *Trichoderma harzianum* in mitigating NaCl stress in Indian mustard (*Brassica juncea* L) through antioxidative defense system". *Frontier in Plant Science*, No. 6: 868.

Ali, RM, Abbas, HM. 2003. "Response of salt stressed barley seedlings to phenylurea". *Plant, Soil and Environment*, No. 49: 158-162.

Anasori, P, Asghari, G. 2008. "Effects of light and differentiation on gingerol and zingiberene production in callus culture of *Zingiber officinale* Rosc". *Research in Pharmaceutical Sciences*, No. 3: 59-63.

Angelova, Z, Georgiev, S, Roos, W. 2006. "Elicitation of plants". *Biotechnology & Biotechnological Equipment*, No. 20: 72-83.

Arehzoo, Z, Sykora, C, Gössnitzer, F, Abrishamchi, P, Asili, J, Mousavi, S, Wawroschd, C. 2015. "Effects of some elicitors on tanshinone production in adventitious root cultures of *Perovskia abrotanoides* Karel". *Industrial Crops and Products*, No. 67: 97-102.

Azhar, N, Hussain, B, Ashraf, M, Abbasi, K. 2011. "Water stress mediated changes in growth: physiology and secondary metabolites of desi ajwain (*Trachyspermum ammi* L.)". *Pakistan Journal of Botany*, No. 43: 15-19.

Bernard YKB, Christie AM, Peebles, JVS, Ka-Yiu, San. 2009. "The effects of UV–B stress on the production of terpenoid indole alkaloids in *Catharanthus roseus* hairy roots". *Biotechnology Progress*, No. 25: 8615.

Bor, M, Seckin, B, Ozgur, R, Yilmaz, O, Ozdemir, F, Turkan, I. 2009. "Comparative effects of drought, salt, heavy metal and heat stresses on gamma-aminobutyric acid levels of sesame (*Sesamum indicum* L.)". *Acta Physiologia Plantarum*, No. 31: 655-659.

Purohit, M, Pande, D, Datta, A, Srivastava, PS. 1995. "Enhanced xanthotoxin content in regenerating cultures of *Ammi majus* and micropropagtion". *Planta Medica*, No. 61: 481-482.

Bourgou, S, Kchouk, M, Bellila, A, Marzouk, B. 2010. "Effect of salinity on phenolic composition and biological activity of *Nigella sativa"*. *Acta Horticulturae*, No. 853: 57–60.

Brachet, J, Cosson, L. 1986. “Changes in the total alkaloid content of *Datura innoxia* Mill. subjected to salt stress”. *Journal of Experimental Botany*, No. 37: 650-656.

Brittain, C, Kremen, C, Garber, A, Klein, A. 2014. “Pollination and Plant Resources Change the Nutritional Quality of Almonds for Human Health”. *PLoS One*, No. 9.2: e90082.

Brown, GD. 2010. “The biosynthesis of artemisinin (Qinghaosu) and the phytochemistry of *Artemisia annua* L. (Qinghao)”. *Molecules*, No. 15: 7603-7698.

Bui, EN. 2013. “Soil salinity: a neglected factor in plant ecology and biogeography”. *Journal of Arid Environment,* No. 92, 14-25.

Cai, Z, Kastell, A, Speiser, C, Smetabska, I. 2013. “Enhanced resveratrol production in *Vitis vinifera* cell suspension cultures by heavy metals without loss of cell viability”. *Applied Biochemistry and Biotechnology*, No. 71: 330-340.

Chadni, Z, Rahaman, MH, Jerin, I, Hoque, KMF, Reza, MA. 2017. “Extraction and optimization of red pigment production as secondary metabolites from *Talaromyces verruculosus* and its potential use in textile industries. Mycology”. *An International Journal on Fungal Biology*, No. 8(1): 48-57.

Chan, LK, See, KS, Bhatt, A. 2010. “Effects of abiotic stress on biomass and anthocyanin production in cell cultures of *Melastoma malabathricum*”. *Biological Research*, No. 43: 127-135.

Chaudhary, P, Jain, D, Anand, K, Mitra, D. “Biofertilizers: a sustainable approach for plant and soil health”. In *Microbial Resources for Sustainable Agriculture*, edited by Debasis Mitra and Divya Jain. LAMBERT Academic Publishing, 2019.

Chen, Y, Guo, Q, Liu, L, Liao, L, Zhu, Z. 2011. “Influence of fertilization and drought stress on the growth and production of secondary metabolites in *Prunella vulgaris* L”. *Journal of Medicinal Plants Research*, No. 5: 1749-1755.

Cho, Y, Njiti, VN, Chen, X, Lighfoot, DA, Wood, AJ. 2003. “Trigonelline concentration in field-grown soybean in response to irrigation”. *Biologia Plantarum*, No. 46: 405-410.

Cho, Y, Lightfoot, DA, Wood, AJ. 1999. “Trigonelline concentrations in salt stressed leaves of cultivated *Glycine max*”. *Phytochemistry*, No. 52: 1235-1238.

Christiansen, JL, Joornsgard, B, Buskov, S, Olsen, CE. 1997. “Effect of drought stress on content and composition of seed alkaloids in narrow-leafed lupin, *Lupinus angustifolius* L”. *European Journal of Agronomy*, No. 7: 307-314.

de Abreu, IN, Mazzafera, P. 2005. "Effect of water and temperature stress on the content of active constituents of *Hypericum brasiliense* Choisy". *Plant Physiology and Biochemistry*, No. 43: 241-248.

de Castro, EM, Pinto, JEBP, Bertolucci, SKV, Malta, MR, Cardoso, MG, Silva, FAM. 2006. "Coumarin contents in young *Mikania glomerata* plants (Guaco) under different radiation levels and photoperiod". *Acta Farmaceutica Bonaerense*, No. 25: 387-392.

Del Moral, R. 1972. "On the variability of chlorogenic acid concentration". *Acta Oecologica*, No. 9: 289-300.

Dixon, R, Paiva, NL. 1995. "Stress–induced phenylpropanoid metabolism". *Plant Cell*, No. 7: 1085-1097.

Emamverdian, A, Ding, Y, Mokhberdoran, F, Xie, Y. 2015. "Heavy metal stress and some mechanisms of plant defense response". *Science World Journal*, No. 2015: 756120.

Fahad, S, Bajwa, AA, Nazir, U, Anjum, SA, Farooq, A, Zohaib, A, Sadia, S, Nasim, W, Adkins, S, Saud, S, Ihsan, MZ, Alharby, H, Wu, C, Wang, D, Huang, J. 2017. "Crop production under drought and heat stress: plant responses and management options". *Frontiers in Plant Science*, No. 8: 1147.

Fatima, S, Mujib, A, Tonk, D. 2015. "NaCl amendment improves vinblastine and vincristine synthesis in *Catharanthus roseus*: a case of stress signaling as evidenced by antioxidant enzymes activities". *Plant Cell Tissue and Organ Culture*, No. 121: 445–458.

Fett–Neto, AG, Pennington, JJ, DiCosmo, F. 1995. "Effect of white light on taxol and baccatin III accumulation in cell cultures of *Taxus cuspidata* Sieb and Zucc". *Journal of Plant Physiology*, No. 146: 584-590.

Fischbach, RJ, Kossmann, B, Panten, H, Steinbrecher, R, Heller, W. 1999. "Seasonal accumulation of Ultraviolet-B screening pigments in needles of Norway spruce (*Picea abies* (L.) Karst)". *Plant, Cell and Environment*, No. 22: 27-37.

Fournier, AR, Proctor, JTA, Gauthier, L, Khanizadeh, S, Andre, B, Andre G, Martine, D. 2003. "Understory light and root ginsenosides in forest-grown *Panax quinquefolius*". *Phytochemistry*, No. 63: 777-782.

Furze, JM, Rhodes, MJC, Parr, AJ, Robins, RJ, Withehead, IM, Threlfall, DR. 1991. "Abiotic factors elicit sesquiterpenoid phytoalexin production but not alkaloid production in transformed root cultures of *Datura stramonium*". *Plant Cell Reports*, No. 10 (3): 111-114.

Ganzera, M, Guggenberger, M, Stuppner, H, Zidorn, C. 2008. "Altitudinal variation of secondary metabolite profiles in flowering heads of *Matricaria chamomilla* cv BONA". *Planta Medica*, No. 74: 453-457.

Gonçalves, S, Romano, A. "Production of plant secondary metabolites by using biotechnological tools". In Secondary metabolites-sources and applications, edited by Vijayakumar Ramasamy and Raja Suresh SS. IntechOpen, 2018.

González-Minero, FJ, Bravo-Diaz, L. 2018. "The use of plants in skin-care products, cosmetics and fragrances: past and present". *Cosmetics*, No. 5: 50.

Grass, S, Zidorn, C, Blattner, FR, Stuppner, H. 2006. "Comparative molecular and phytochemical investigation of *Leontodon autumnalis* (Asteraceae, Lactuceae) populations from central Europe". *Phytochemistry*, No. 67: 122-131.

Griffith, M, Yaish, MWF. 2004. "Antifreeze proteins in overwintering plants: a tale of two activities". *Trends in Plant Science*, No. 9: 399-405.

Groppa, MD, Tomaro, ML, Benavides, MP. 2001. "Polyamines as protectors against cadmium or copper-induced oxidative damage in sunflower leaf discs". *Plant Science*, No. 161: 481-488.

Gull, A, Ahmad, LA, Islam, NUI. "Biotic and abiotic stress in plants". In *Abiotic and Biotic Stress in Plants*, edited by de Oliveira Alexandre Bosco. IntechOpen, 2019.

Haghighi, Z, Karimi, N, Modarresi, M, Mollaei, S. 2012. "Enhancement of compatible solute and secondary metabolites production in Plantago ovate Forsk by salinity stress". *Journal of Medicinal Plants Research*, No. 6: 3495-3500.

Hagimori, M, Matsumoto, T, Obi, Y. 1982. "Studies on the production of *Digitalis cardenolides* by plant tissue culture III. Effects of nutrients on digitoxin formation by shoot–forming cultures of *Digitalis purpurea* L. grown in liquid media". *Plant and Cell Physiology*, No. 23: 1205-1211.

Hernáendez, I, Alegre, L, Munne-Bosch, S. 2006. "Enhanced oxidation of flavan-3-ols and proanthocyanidins accumulation in water-stressed tea plants". *Phytochemistry*, No. 67: 1120-1126.

Huang, ZA, Zhao, T, Fan, HJ, Ning, W, Zheng, SS, Ling, HQ. 2012. "The up-regulation of NtaN2 expression at low temperature is required for anthocyanin accumulation in juvenile leaves of lc-transgenic tobacco (*Nicotiana tabacum* L.)". *Journal of Genetics and Genomics*, No. 20: 149-156.

Hussain, MS, Fareed, S, Rahman, MA, Ansari, S. 2012. "Current approaches towards production of secondary plant metabolites". *Journal of Pharmacy and Bioallied Sciences*, No. 4(1): 10-20.

Hussein, RA, El- Anssary, AA. "Plants secondary metabolites: the key drivers of the pharmacological actions of medicinal plants". In *Herbal Medicine*, edited by Philip F. Builders. IntechOpen, 2018.

Isah, T. 2019. "Stress and defense responses in plant secondary metabolites production". *Biological Research*, No. 52.

Jaakola, L, Maatta- Riihinen, K, Karenlampi, S, Hohtola, A. 2004. "Activation of flavonoid biosynthesis by solar radiation in bilberry (*Vaccinium myrtillus* L.) leaves". *Planta*, No. 218: 721-728.

Jabr, F, 2013. "Farming a Toxin to Keep Crops Healthy". *Scientific American*.

Jensen, CR, Mongensen, VO, Mortensen, G, Fieldsend, JK, Milford, GFJ, Andersen, MN, Thage, JH. 1996. "Seed glucosinolate, oil and protein contents of field-grown rape *(Brassica napus* L.) affected by soil drying and evaporative demand". *Field Crops Research*, No. 47: 93-105.

Jiang, QY, Zhuo, F, Long, SH, Zhao, HD, Yang, DJ, Ye, ZH, Li, SS, Jing, YX. 2016. "Can arbuscular mycorrhizal fungi reduce Cd uptake and alleviate Cd toxicity of *Lonicera japonica* grown in Cd-added soils?". *Science Reporter*, No. 6: 21805.

Jochum, GM, Mudge, KW, Thomas, RB. 2007. "Elevated temperatures increase leaf senescence and root secondary metabolite concentrations in the understory herb *Panax quinquefolius* (Araliaceae)". *American Journal of Botany*, No. 94: 819-826.

Kabera, J, Semana, E, Mussa, AR, He, X. 2014. "Plant secondary metabolites: biosynthesis, classification, function and pharmacological properties". *Journal of Pharmacy and Pharmacology*, No. 2(2014): 377-392.

Kakegawa, K, Hattori, E, Koike, K, Takeda, K. 1991. "Induction of anthocyanin synthesis and related enzyme activities in cell cultures of *Centaurea cyanus* by UV-light irradiation". *Phytochemistry*, No. 30: 2271-2273.

Kallscheuer, N, Classen, T, Drepper, T, Marienhagen, J. 2018. "Production of plant metabolites with applications in the food industry using engineered microorganism". *Current Opinion in Biotechnology*, No. 56: 7-17.

Kennedy, BF, Filippis, LFD. 1999. "Physiological and oxidative response to NaCl of the salt tolerant *Grevillea ilicifolia* and the salt sensitive *Grevillea Arenaria*". *Journal of Plant Physiology*, No. 155: 746-754.

Khalid, A. 2006. "Influence of water stress on growth essential oil, and chemical composition of herbs (*Ocimum* sp.)". *International Agrophysics*, No. 20: 289-296.

Kliebenstein, D. 2012. "Making new molecules-evolution of structures for novel metabolites in plants". *Current Opinion in Plant Biology*, No. 16, 1-6.

Kovácik, J, Klejdus, B, Hedbavny, J, Backor, M. 2009. "Salicylic acid alleviates NaCl-induced changes in the metabolism of *Matricaria chamomilla* plants". *Ecotoxicology*, No. 18: 544-554.

Kramer, GF, Norman, HA, Krizek, DT, Mirecki, RM. 1991. "Influence of UV-B radiation on polyamines, lipid peroxidation and membrane lipids in cucumber". *Phytochemistry*, No. 30: 2101-2108.

Krishnamurthy, R, Bhagwat, KA. 1989. "Polyamines as modulators of salt tolerance in rice cultivars". *Plant Physiology*, No. 91: 500-504.

Ksouri, R, Megdiche, W, Debez, A, Falleh, H, Grignon, C, Abdelly, C. 2007. "Salinity effects on polyphenol content and antioxidant activities in leaves of the halophyte *Cakile maritima*". *Plant Physiology and Biochemistry*, No. 45: 244-249.

Kumar, A, Singhal, KC, Sharma, RA, Vyas, GK, Kumar, V. 2013. "Total phenolic and antioxidant activity of *Catharanthus roseus* in different geographical locations of Rajasthan". *Asian Journal of Experimental Biological Science*, No. 4: 155-158.

Kumar, I, Sharma, RK. 2018. "Production of secondary metabolites in plants under abiotic stress: an overview". *Significance of Bioengineering and Bioscience*, No. 2(4).

Lavoir, AV, Staudt, M, Schnitzler, JP, Landais, D, Massol, F, Rocheteau, A, Rodriguez, R, Zimmer, I, Rambal, S. 2009. "Drought reduced monoterpene emissions from *Quercus ilex* trees: results from a throughfall displacement experiment within a forest ecosystem". *Biogeosciences*, No. 6: 863-893.

Lee, KT, Yamakawa, T, Kodama, T, Shimomura, K. 1998. "Effects of chemicals on alkaloid production by transformed roots of *Atropa belladonna*". *Phytochemistry*, No. 49 (8): 2343-2347.

Li, B, Wang, B, Li, H, Peng, L, Ru, M, Liang, Z, Yan, X, Zhu, Y. 2016. "Establishment of *Salvia castanea* Diels f. tomentosa Stib. hairy root cultures and the promotion of tanshinone accumulation and gene expression with Ag+, methyl jasmonate, and yeast extract elicitation". *Protoplasma*, No. 253 (1): 87-100.

Li, WD, Hou, JL, Wang, WQ, Tang, XM, Liu, CL, Xing, D. 2011. "Effect of water deficit on biomass production and accumulation of secondary metabolites in roots of *Glycyrrhiza uralensis*". *Russian Journal of Plant Physiology*, No. 58: 538–542.

Liu, CZ, Guo, C, Wang, YC, Ouyang, F. 2002. "Effect of light irradiation on hairy root growth and artemisinin biosynthesis of *Artemisia annua* L". *Process Biochemistry*, No. 38: 581-585.

Liu, H, Wang, X, Wang, D, Zou, Z, Liang, Z. 2011. "Effect of drought stress on growth and accumulation of active constituents in *Salvia miltiorrhiza* Bunge". *Industrial Crops and Products*, No. 33: 84-88.

Liu, L, Gitz III, DC, McClure, JW. 1995. "Effects of UV-B on flavonoids, ferulic acid, growth and photosynthesis in barley primary leaves". *Physiologia Plantarum*, No. 93: 734-738.

Liu, W, Liu, C, Yang, C, Wang, L, Li, S. 2010. "Effect of grape genotype and tissue type on callus growth and production of resveratrols and their piceids after UV–C irradiation". *Food Chemistry*, No. 122: 475-481.

Lobay, D. 2015. "Rauwolfia in the treatment of hypertension". *Integrative Medicine: A Clinician's Journal*, No. 14(3): 40-46.

Mader, J. 1999. "Effects of jasmonic acid, silver nitrate and L-AOPP on the distribution of free and conjugated polyamines in roots and shoots of *Solanum tuberosum, in vitro*". *Journal of Plant Physiology*, No. 154: 79-88.

Misra, N, Gupta, AK. 2006. "Effect of salinity and different nitrogen sources on the activity of antioxidant enzymes and indole alkaloid content in *Catharanthus roseus* seedlings". *Journal of Plant Physiology*, No. 163 (1): 11-18.

Mizukami, H, Konoshima, M, Tabata, M. 1977. "Effect of nutritional factors on shikonin derivative formation in Lithospermum callus cultures". *Phytochemistry*, No. 16: 1183-1186.

Nadeau, P, Delaney, S, Chouinard, L. 1987. "Effects of cold hardening on the regulation of polyamine levels in wheat (*Triticum aestivum* L.) and Alfalfa (*Medicago sativa* L.)". *Plant Physiology*, No. 8: 73-77.

Namdeo, AG, Sharma, A, Fulzele, DP, Mahadik, KR. 2010. "Influence of geographical and climatic conditions on camptothecin content of *Nothapodytes nimmoniana*". *Records of Natural Products*, No. 4: 64-71.

Narula, A, Kumar, S, Srivastava, PS. 2005. "Abiotic metal stress enhances diosgenin yield in *Dioscorea bulbifera* L. cultures". *Plant Cell Reports*, No. 24: 250–254.

Obrenović, S. 1990. "Effect of Cu (11) D–penicillanine on phytochrome mediated betacyanin formation in *Amaranthus caudatus* seedlings". *Plant Physiology and Biochemistry*, No. 28: 639-646.

Ohlsson, AB, Berglund, T. 1989. "Effect of high MnSO4 levels on cardenolide accumulation by *Digitalis lanata* tissue cultures in light and darkness". *Journal of Plant Physiology*, No. 135: 505-507.

Pagare, S, Bhatia, M, Tripathi, N, Pagare, S, Bansal, YK. 2015. "Secondary metabolites of plants and their role: overview". *Current Trends in Biotechnology and Pharmacy*, No. 9(3): 293-304.

Pareek, A, Sopory, S, Bohnert, HJ, Govindjee. 2010. Abiotic Stress Adaptation in Plants: Physiological, Molecular and Genomic Foundation. Dordrecht: Springer, 526.

Parsaeimehr, A, Sargsyan, E, Vardanyan, A. 2011. "Expression of secondary metabolites in plants and their useful perspective in animal health. Animal Biology & Animal Husbandry". *International Journal of the Bioflux Society*, No. 3(2): 115-124.

Pedranzani, H, Racagni, G, Alemano, S, Miersch, O, Ramirez, I, Pena-Cortes, H, Taleisnik, E, Machado-Domenech, E, Abdala, G. 2003. "Salt tolerant tomato plants show increased levels of jasmonic acid". *Plant Growth Regulation*, No. 412: 149-158.

Gottlieb, LD. 1988. Plant Evolutionary Biology. Springer Science + Business Media B.V.

Queslati, S, Karray-Bouraoui, N, Attia, H, Rabhi, M, Ksouri, R, Lachaal, M. 2010. "Physiological and antioxidant responses of *Mentha pulegium* (Pennyroyal) to salt stress". *Acta Physiologiae Plant*, No. 32: 289-296.

Radusiene, J, Karpaviciene, B, Stanius, Z. 2012. "Effect of external and internal factors on secondary metabolites accumulation in St. John's wort". *Botanica Lithuanic*a, No. 18: 101–108.

Ramakrishna, A, Ravishankar, GA. 2011. "Influence of abiotic stress signals on secondary metabolites in plants". *Plant Signaling and Behavior*, No. 6(11): 1720-1731.

Razmjoo, K, Heydarizadeh, P, Sabzalian, MR. 2008. "Effect of salinity and drought stresses on growth parameters and essential oil content of *Matricaria chamomilla*". *International Journal of Agriculture and Biology*, No. 10: 451-454.

Saba, S, Katare, DP, Iqbal, M, Srivastava, PS. 2000. "Effect of ZnSO4 and CuSO4 on regeneration and lepidine content in *Lepidium sativum*". *Biologia Plantarum*, No. 43: 253-256.

Said-Al Ahl, HAH, Omer, EA. 2011. "Medicinal and aromatic plants production under salt stress. A review". *Herba Polonica Journal*, No. 57: 72-87.

Sampaio, B. L., Edrada-Ebel, R., Da Costa, F. B. 2016. "Effect of the environment on the secondary metabolic profile of *Tithonia diversifolia*: a model for environmental metabolomics of plants". *Scientific Reports*, No. 6 (2016): 29265.

Savitha, B. C., Thimmaraju, R., Bhagyalakshmi, N, Ravishankar, G. A. 2006. "Different biotic and abiotic elicitors influence betalain production in hairy root cultures of *Beta vulgaris* in shake-flask and bioreactor". *Process Biochemistry*, No. 41: 50-60.

Seca, AML, Pinto, DCGA. 2018. "Plant secondary metabolites as anticancer agents: successes in clinical trials and therapeutic application". *International Journal of Molecular Science*, No. 19(1): 263.

Shakeran, Z, Keyhanfar, M, Asghari, G, Ghanadian, M. 2015. "Improvement of atropine production by different biotic and abiotic elicitors in hairy root cultures of *Datura metel*". *Turkish Journal of Biology*, No. 39: 111–118.

Sharma, M, Ahuja, A, Gupta, R, Mallubhotla, S. 2015. "Enhanced bacoside production in shoot cultures of *Bacopa monnieri* under the influence of abiotic elicitors". *Natural Product Research*, No. 29 (8): 745-749.

Shiga, T, Shoji, K, Shimada, H, Hashida, SN, Goto, F, Yoshihara, T. 2009. "Effect of light quality on rosmarinic acid content and antioxidant activity of sweet basil *Ocimum basilicum* L". *Plant Biotechnology Journal*, No. 26: 255-259.

Singh, S, Sinha, S. 2005. "Accumulation of metals and its effects in *Brassica juncea* (L.) Czern. (cv. Rohini) grown on various amendments of tannery waste". *Ecotoxicology and Environmental Safety*, No. 62: 118-127.

Szabo, B, Tyihak, E, Szabo, G, Botz, L. 2003. "Mycotoxin and drought stress induced change of alkaloid content of *Papaver somniferum* plantlets". *Acta Botanica Hungarica*, No. 45: 409-417.

Tari, I, Kiss, G, Deer, AK, Csiszar, J, Erdei, L, Galle, A, Gemes, K, Horvath, F, Poor, P, Szepesi, A, Simon, LM. 2010. "Salicylic acid increased aldose reductase activity and sorbitol accumulation in tomato plants under salt stress". *Biologia Plantarum*, No. 54: 677-683.

Thirumurugan, D, Cholarajan, A, Raja, SSS, Vijayakumar, R. 2018. "An introductory chapter: secondary metabolites". In *Secondary Metabolite, Secondary Metabolites: Sources and Applications*, edited by Vijayakumar Ramasamy and Raja Suresh S.S., IntechOpen.

Threlfall, DR, Whitehead, IM. 1988. "The use of biotic and abiotic elicitors to induce the formation of secondary plant products in cell suspension cultures of Solanaceous plants". *Biochemical Society Transactions*, No. 16: 71-75.

Tiwari, R, Rana, CS. 2015. "Plant secondary metabolites: a review". *International Journal of Engineering Research and General science*, No. 3(5): 661-670.

Trejo-Tapia, G, Jimenez- Aparicio, A, Rodriguez-Monroy, M, Jesus-Sanchez, A D, Gutierrez-Lopez, G. 2001. "Influence of cobalt and other microelements on the production of betalains and the growth of suspension cultures of *Beta vulgaris*". *Plant Cell Tissue and Organ Culture*, No. 67: 19-23.

Varshney KA, Gangwar, LP, Goel, N. 1988. "Choline and betaine accumulation in *Trifolium alexandrinum* L. during salt stress". *Egyptian Journal of Botany*, No. 31: 81-86.

Vats, S. 2018. "Larvicidal activity and in vitro regulation of rotenoids from *Cassia tora* L". 3 *Biotech*, No. 8(1): 13.

Wang, W, Tang, K, Yang, HR, Wen, PF, Zhang, P, Wang, HL, Huang, WD. 2010. "Distribution of resveratrol and stilbene synthase in young grape plants (*Vitis vinifera* L. cv. Caber- net Sauvignon) and the effect of UV–C on its accumulation". *Plant Physiology and Biochemistry*, No. 48: 142-152.

War, AR, Paulraj, MG, Ahmad, T, Buhroo, AA, Hussain, B, Ignacimuthu, S, Sharma, HC. 2012. "Mechanism of plant defense against insect herbivore". *Plant signaling and Behavior*, No. 7(10): 1306-1320.

Xu, Z, Jiang, Y, Jia, B, Zhou, G. 2016. "Elevated-CO2 response of stomata and its dependence on environmental factors". *Frontier in Plant Science*, No. 7: 657.

Yang, L, Wen, KS, Ruan, X, Zhao, YX, Wei, F, Wang, Q. 2018. "Response of Plant Secondary Metabolites to Environmental Factors". *Molecules*, No. 23(4): 762.

Yu, KW, Murthy, HN, Hahn EJ, Paek, KY. 2005. "Ginsenoside production by hairy root cultures of *Panax ginseng*: influence of temperature and light quality". *Biochemical Engineering Journal*. No. 23: 53-56.

Zhang, L, Wang, Q, Guo, Q, Chang, Q, Zhu, Z, Liu, L, Xu, H. 2012. "Growth, physiological characteristics and total flavonoid content of *Glechoma longituba* in response to water stress". *Journal of Medicinal Plants Research*, No. 6: 1015-1024.

Zhao, Y, Qi, LW, Wang, WM, Saxena, PK, Liu, CZ. 2011. "Melatonin improves the survival of cryopreserved callus of *Rhodiola crenulate*". *Journal of Pineal Research*, No. 50: 83-88.

Zhi-lin, Y, Chuan-Chao, D, Lian-qing, C. 2007. "Regulation and accumulation of secondary metabolites in plant-fungus symbiotic system". *African Journal of Biotechnology*, No. 6: 1266-1271.

Zidorn, C, Stuppner, H. 2001. "Evaluation of chemosystematics characters in the genus Leontodon". *Taxon*, No. 50: 115-133.

Zobayed, SMA, Afreen, F, Kozai, T. 2005. "Temperature stress can alter the photosynthetic efficiency and secondary metabolite concentrations in St. John's wort". *Plant Physiology and Biochemistry*, No. 43: 977-984.

Zobayed, SMA, Afreen, F, Kozai, T. 2007. "Phytochemical and physiological changes in the leaves of St. John's wort plants under a water stress condition". *Environmental and Experimental Botany,* No. 59: 109-116.

CHAPTER TWO

THE CONSEQUENCES OF OXIDATIVE STRESS IN PLANTS

G GARG, V GUPTA AND S BHATI

Introduction

Reactive derivatives of molecular oxygen

When plants are exposed to biotic and abiotic stresses, they undergo a complex chemical and physiological phenomenon resulting in oxidative stress and producing oxygen molecules called ‘reactive oxygen species’ (ROS). In the ground state, O_2 is a totally harmless molecule; it has two unpaired electrons with a parallel spin, which makes it paramagnetic. When a plant is exposed to environmental stress conditions, these O_2 molecules absorb sufficient energy to activate them as free radicals. O_2 is sequentially reduced to singlet or activated oxygen ($^1O_2$), hydrogen peroxide (H_2O_2), hydroxyl ions (OH^-), activated oxygen (O_2), and superoxide anions (O_2^-), which all play pivotal roles in plant metabolism. Besides this, there are other forms of ROS involved, including peroxyl, alkoxyl, and hydroperoxyl radicals, peroxynitrite, ozone, and hypochlorous acid. In healthy cells, ROS are generated in a controlled manner or there is no generation; however, under conditions of oxidative stress, ROS production increases dramatically, resulting in subsequent alteration to membrane lipids, proteins, and nucleic acids. Superoxides, peroxyl radicals, hydrogen peroxide, hydroxyl radicals, and peroxynitrite are released and react with nucleic acids, proteins, and lipids, resulting in damage to cells and tissues. ROS are major initiators of lipid peroxidation and membrane-bound polyunsaturated fatty acids like arachidonic acid and linoleic acid are their major targets, resulting in damage to cell membranes.

Relationship between environmental stress and ROS

The cellular compartments (the chloroplast, mitochondria, peroxisome, and cytoplasm) of plants are involved in the generation of ROS inside cells. Under conditions of optimal growth, intracellular ROS are mainly produced at low levels in the organelles. Abiotic stresses, such as those caused by drought, salinity, and low or high temperatures, often induce stomatal closure, which limits CO_2 uptake. Generation of $NADP^+$ through the Calvin cycle is reduced and photo-respiratory production of H_2O_2 in the peroxisome and production of superoxide and H_2O_2 or singlet oxygen is favoured by the reduced photosynthetic electron transport (Apel and Hirt, 2004).

In addition to the organelles, the plasma membrane, being a selective barrier between living cells and their environments, plays a fundamental role in responding to external changes. Because of their ability to rapidly modify their physical state, cellular membranes are not only the primary sites of stress damage, but also able to recognize environmental insult and activate stress-responsive-defense genes. The micro-domain organization and physical state of the cell membranes allows for sensitive monitoring of changes in the environmental conditions. Ion channels, one component of the cell membranes, play a vital role in the transduction of environmental and internal signals. Ion channels help facilitate ionic exchange through hydrophilic pathways. They function as permeation pores through which electrically charged ion species can pass across biological membranes. It has been reported that ion channels are directly involved in important cellular processes, such as plant defense responses induced by elicitors, light perception, and mechanical signals [**11a, b**].

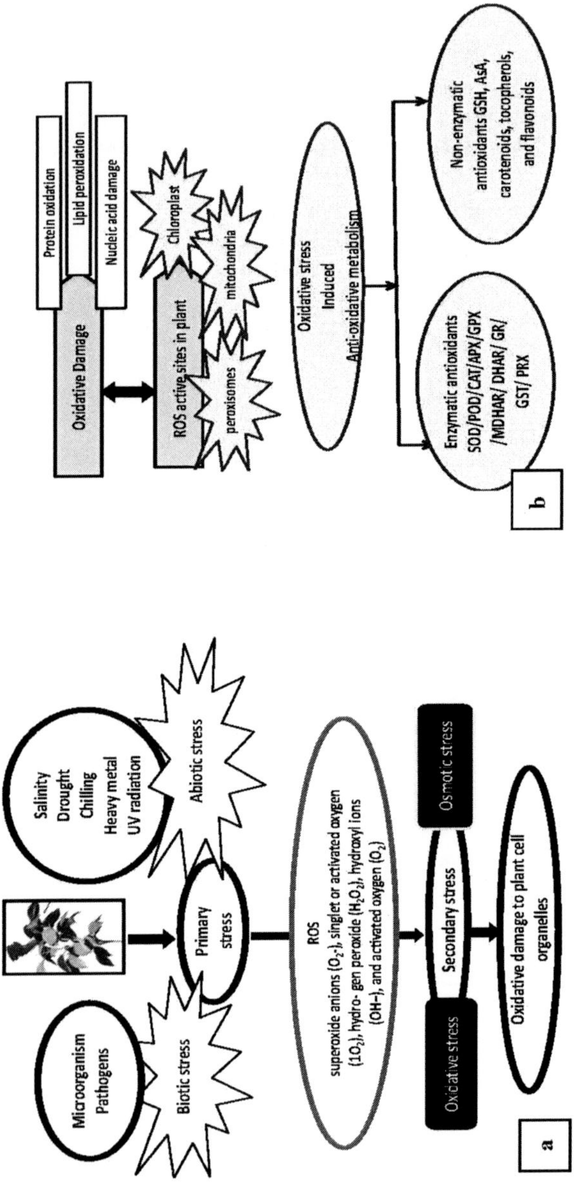

11 (a, b). Biological source of reactive oxygen in plants and ROS.

The plasma membrane, together with the apoplast, is the main site of ROS generation in response to endogenous signals and exogenous environmental stimuli. Several types of enzymes, such as NADPH oxidases, amine oxidases, polyamine oxidases, oxalate oxidases, and a large family of class III peroxidases, which are localized at the membrane-cell surface or apoplast, contribute to the production of apoplastic ROS (Apel and Hirt, 2004; Cosio and Dunand, 2009; Gill and Tuteja, 2010). During exposure to stress, the major role of phospholipids, the backbone of cellular membranes, may be to serve as precursors for the generation of secondary messenger signaling molecules, such as phosphotidylinositol, inositol-1,4,5-tiphosphate (IP3), diacylglycerol, and jasmonates. IP3 and diacylglycerol are secondary messengers that can activate a protein kinase and induce Ca^{2+} release, respectively. Additionally, IP3 itself is a signal molecule and may be involved in several processes, such as the recruitment of signaling complexes to specific membrane locations and their assembly. In addition, under oxidative stress, polyunsaturated fatty acids (PUFAs) are attacked by different ROS, particularly $^1O_2$ and $\cdot OH$. This causes the production of lipid hydroperoxides, leading to a reduction in membrane fluidity. In turn, this change in the membrane's physical state can activate downstream signaling intermediates. Overproduction of highly reactive ROS caused by stress conditions in plant cells is toxic to the proteins, lipids, and nucleic acid of the plant's cells, ultimately resulting in cellular damage and their death (Gill and Tuteja, 2010). On the other hand, increased production of ROS during stress conditions is also thought to signal the activation of stress response pathways (Baxter et al., 2014). Plants have evolved efficient enzymatic and non-enzymatic antioxidative systems to protect themselves against oxidative damage and fine modulation of low levels of ROS for signal transduction. ROS-scavenging enzymes in plants include superoxide dismutase (SOD), ascorbate peroxidase (APX), catalase (CAT), glutathione peroxidase (GPX), monodehydroascorbate reductase (MDHAR), dehydroascorbate reductase (DHAR), glutathione reductase (GR), glutathione S-transferase (GST), and peroxiredoxin (PRX). These antioxidant enzymes are located in different sites in the plant cells and work together to detoxify RO: SOD acts as the first line of defense converting O_2^- into H_2O_2. It is followed by CAT, APX, and GPX in detoxifying toxic levels of H_2O_2. In contrast to CAT, APX requires an ascorbic acid (AsA) and/or a glutathione (GSH) regenerating cycle involving MDHAR, DHAR, and GR. GPX, GST, and PRX reduce H_2O_2 and organic hydroperoxides through ascorbate-independent thiol mediated pathways using GSH, thio-redoxin (TRX) or glutaredoxin (GRX) as a nucleophile (Dietz et al., 2006; Meyer et al., 2012; Noctor et al., 2014).

Non-enzymatic antioxidants, including GSH, AsA, carotenoids, tocopherols, and flavonoids, are also crucial for ROS homeostasis in plants (Gill and Tuteja, 2010). Besides traditional enzymatic and non-enzymatic antioxidants, increasing evidence indicates that soluble sugars, including disaccharides, raffino-oligosaccharides, and fructans, have a dual role with respect to ROS (Couee et al., 2006; Keunen et al., 2013). Soluble sugars are directly associated with the production rates of ROS via ROS regulated-metabolic pathways, such as mitochondrial-respiration or photosynthesis. Conversely, they also feed the NADPH-producing metabolism to participate in antioxidative processes (Couee et al., 2006). Alternative oxidases (AOX) also prevent the excess generation of ROS in the electron transport chains of mitochondria (Maxwell et al., 1999). By diverting electrons flowing through electron transport chains, AOX can decrease the risk of electron leakage to O_2, helping in the generation of O_2^-.

Location of ROS formation in plant cellular organelles under stress conditions

Biotic and abiotic factors, including salinity, drought, high and low temperatures, metal toxicity, pollutants, ultraviolet light, and high doses of insecticides or pesticides, as well as pathogen infection, lead to oxidative stress in plants (Foyer, 1997; Del et al., 2006; Blokhina and Fagersted, 2010; Shah et al., 2001). This oxidative stress is caused, either directly or indirectly, in both unstressed and stressed cells at several locations in the chloroplasts, mitochondria, plasma membranes, peroxisomes, apoplast, endoplasmic reticulum, and cell walls, etc., leading to ROS accumulation. The plant cells are in a state of oxidative stress if the quantity of ROS is greater than the internal defense mechanisms can handle. This inhibits seed germination (Han, 2009), polar cell growth (Maheshwari and Dubey, 2009), lignin biosynthesis in the cell wall (Tanou et al., 2009), and, finally, retards plant growth via flower and leaf abscission (Sharma and Dubey, 2005), including production of ROS [**11a, b**]. ROS are formed when there is a leakage of electrons onto O_2 from the electron transport activities of the chloroplasts, mitochondria and plasma membranes of various metabolic pathways localized in the different cellular compartments. In the mitochondria and chloroplasts, different types of abiotic stresses, such as drought, salt, and temperature, as well as combinations of these conditions with high-light stress, enhance the production of ROS. In both cell organelles, electron transport chain compartments (ETCs) are the main site of ROS synthesis. The photosystems of the chloroplasts (PSI and PSII) are key regulatory sites, responsible for the synthesis of ROS; whereas in mitochondria the direct reduction of oxygen to $O^{2\bullet-}$ occurs in the flavor-protein region of NADH dehydrogenase segment of complex I and the ubiquinone-cytochrome region (complex III) of the ETC (Arora et al., 2002).

In the endoplasmic reticulum, NADPH dependent electron transport involving Cyt-P450 produces $O^{2\bullet-}$ (Mittler, 2002). Peroxisomes are probably the major sites of intracellular H_2O_2 production due to their oxidative-metabolic pathway (Del Rio et al., 2006). Using electron spin resonance spectroscopy (ESR), two important sites have been identified. The organelle matrix and peroxisomal membranes, with molecular weights of 18 and 29 kDa respectively, and the integral peroxisomal membrane polypeptides (PMPs), with a molecular mass of 32 kDa, are the main sites for ROS synthesis in the peroxisome. The main metabolic processes responsible for the generation of H_2O_2 in different types of peroxisomes are: the glycolate oxidase reaction (photorespiration); β-oxidation of fatty acid; and the enzymatic reaction of flavin oxidases.

In the plasma membranes, electron transporting oxidoreductases constitute an ever-present site for the generation of ROS. Heyno et al. (2011) isolated the plasma membrane from the growing and the non-growing zones of hypocotyls and roots of etiolated soybean seedlings, along with the coleoptiles and roots of etiolated maize seedlings. They studied the production of ROS with the help of ESR techniques and finally concluded that NADPH mediates the production of $O^{2\bullet-}$ in all plasma membranes.

Apart from the major cell organelles, the plant cell wall plays a crucial role in maintaining redox balance in the cell. Peroxidase (POX) in the cell wall has shown hypersensitive reactions during the production of ROS (Gross, 1977; Martinez, 1998). Oxidative bursting can occur in the plant cell wall where molecular oxygen is reduced to $O^{2\bullet-}$ and at acidic pH values this active form of ROS causes spontaneous mutation. Class III POX present in the cell wall will oxidize NADH and catalyze the formation of $O^{2\bullet-}$. Sometimes, monophenols, especially of coniferyl alcohol, malate dehydrogenase, and diamine oxidases, are involved in the production of activated oxygen in the cell wall. Enzymes located in the cell wall, oxalate oxidase (also known as germin), and amine oxidase etc., have been shown to be responsible for apoplastic ROS production (Apel and Hirt, 2004).

Correlation between oxidative stress and biotic and abiotic stress

Under normal growth conditions, there is low level of production of ROS in plants. However, in response to various environmental stresses, ROS can be drastically increased, upsetting the normal balance of $O^{2\bullet-}$, •OH, and H_2O_2 in the intracellular matrix of the cell. Various environmental stresses, such as drought, salinity, chilling, metal toxicity, UV-B radiation, and pathogen attack, etc., alter production of ROS. How oxidative stress affects plant growth and metabolic activity under different stresses is discussed below.

Correlation between drought and oxidative stress

Drought affects the plant–water potential, turgor pressure, and the mass flow movements of nutrients in plant cells. It changes the physiological and morphological traits of a plant. In the first instance, drought affects growth parameters, such as fresh weight and water content. Drought induces free radicals, affecting antioxidant defenses and ROS production, including superoxide radicals, hydrogen peroxide, and hydroxyl radicals, and resulting in oxidative stress. At high concentrations, ROS cause impairment at various levels of the organization of the plant system. They initiate lipid per-oxidation, membrane deterioration, and degradation of important biomolecules like proteins, lipids, and nucleic acids (Heidari, 2011; Sgherri et al., 2000; Nair et al., 2008) and can cause severe damage to the photosynthetic machinery. Drought stress decreases chlorophyll content and increases the photo-oxidation process. A decrease in the chlorophyll content of *Paulownia imperialis* (Foxglove tree), beans, and *Carthamus tinctorius* (Safflower) has been noted under drought conditions (Astorga and Melendez, 2010; Beinsan et al., 2003; Siddiqi et al., 2009). Under drought conditions, CO_2 fixation decreases due to stomatal closure, which, in turn, leads to reduced $NADP^+$ regeneration through the Calvin cycle. Due to the lack of an electron acceptor, too great a reduction in the photosynthetic ETC occurs, which leads to higher leakage of electrons onto O_2 molecules through the Mehler reaction. Biehler and Fock (1996) reported 50 % more leakage of photosynthetic electrons through the Mehler reaction in drought stressed wheat plants when compared to unstressed plants. An excess of light energy in the antenna part of the PS II photosystem leads to the generation of dangerous ROS. Drought stress enhances the photo-respiratory pathway, especially when RUBP oxygenation is maximized due to an inadequacy in CO_2 fixation. Drought affects biochemical activities, like those of nitrate reductase (NR), due to lower uptake of nitrates from the soil. It increases the biosynthesis of ethylene, which restricts plant growth (Ali et al., 2014). Drought is a multi-dimensional stress, affecting various sub-cellular compartments, cell organs, and at the level of the whole plant (Choluj et al., 2004; Rahdari et al., 2012).

Correlation between salinity and oxidative stress

Roots are the first organs to undergo salinity stress, resulting in greater reduction in growth compared to shoots (Bernstein et al., 2004). Moreover, some developmental stages, including germination, seedling emergence, and flowering, are highly susceptible to salt stress. Under salinity stress, excess accretion of Na^+ influences the efflux of cytosolic K^+ and Ca^{2+}, consequently leading to an imbalance in cellular homeostasis, nutrient deficiency, oxidative stress, retarded growth, and cell death. It has been reported in several previous studies that a high level of salinization severely reduces the rate of photosynthesis due to stomatal restriction (stomatal closure) and/or non-stomatal restrictions (chlorophyll malfunctioning/deprivation of enzymatic proteins/destruction of the chloroplastic membrane ultra-structure). It engages cellular electron transport within the different sub-cellular compartments of the chloroplast.

Salinity stress leads to the exposure of the chloroplasts to excessive excitation energy and over-reduction of the photosynthetic electron transport system leading to enhanced generation of the ROS, which induces oxidative stress in plants. This oxidative stress disrupts normal physiological metabolism through lipid peroxidation and the denaturation of proteins and nucleic acids [**12**].

Low temperatures (0–15 °C) affect plant growth; this is also called chilling, (i.e. not stress due to freezing temperatures) (Wang et al., 2016). Different plants have different capacities to tolerate chilling stress. Many tropical and subtropical plants, including tobacco, rice, and maize, cannot survive chilling, whereas *Arabidopsis* and some over-winter cereals continue to grow. A series of physiological and cellular alterations occur in plants during chilling, including alterations to the membrane structure, photosynthesis, calcium signaling, and metabolism. These changes assist plant adaptation, helping them to grow during chilling (Liu et al., 2008).

12. Effects of oxidative stress caused by salinity.

<table>
<tr><td rowspan="8">Salinity stress</td><td rowspan="4">Na⁺ uptake</td><td>Increased ion toxicity</td><td rowspan="8">Oxidative damage</td></tr>
<tr><td>Disintegration of cellular membrane ultra-structure</td></tr>
<tr><td>Leaf expansion inhibition</td></tr>
<tr><td>Early leaf abscission</td></tr>
<tr><td rowspan="4">Physiological destruction</td><td>Stomatal conductance inhibition</td></tr>
<tr><td>Decreased intercellular CO_2 and limited carboxylation</td></tr>
<tr><td>Decline PEPcase, NADP-ME RUBISCO activity</td></tr>
<tr><td>Decline photosynthesis and electron transport</td></tr>
</table>

Correlation between cold and oxidative stress

Chloroplasts are organelles that perform energy transformation in plants and are essential for growth and development. A chilling stress is a common environmental stress that can directly affect the physiological functions of the chloroplasts. First, chilling stress can transform the lipid phase of the chloroplast membrane before affecting the enzymatic regulation of the chloroplasts. Finally, it reduces the photosynthetic efficiency of the plant and sees the production of excess ROS. The accumulation of ROS acts as a signal that initiates the acclimatization process in plants under temperature stress. The acclimatization process can reduce the activity of two key regulatory enzymes of the Calvin cycle, i.e fructose-1,6-bisphosphatase (FBPase) and sedoheptulose-1,7-bisphosphatase (SBPase), which promote plant adaptation to low temperatures by regulating nucleus and chloroplast gene expression. It regenerates damaged proteins and thus regulates lipid homeostasis.

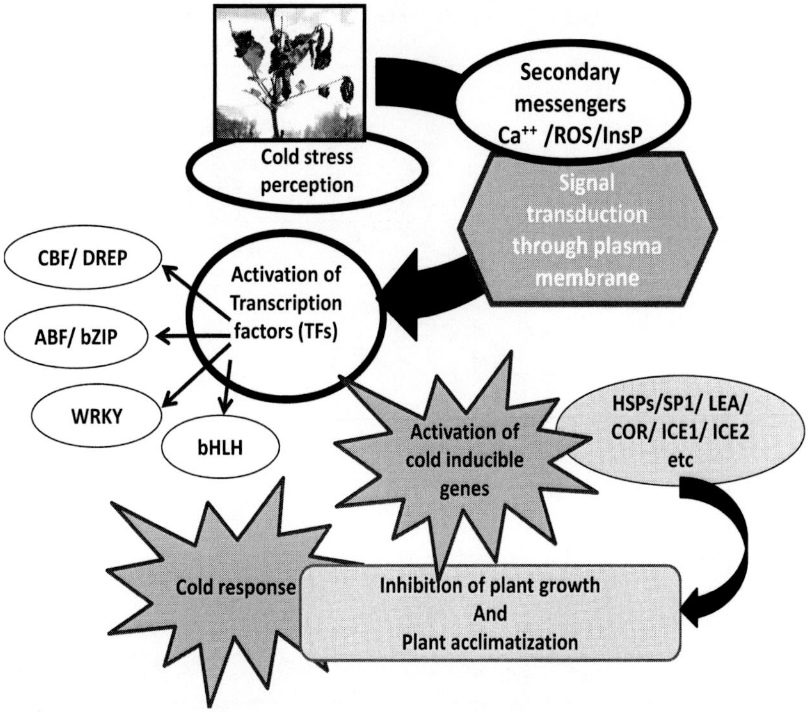

13. A schematic presentation of the transcriptional regulation of transcription factors playing key roles in cold stress conditions in plants.

The lipid phase composition is crucial to the photosynthetic process of the chloroplasts and is closely related to the adaption of the plant to chilling tolerance. The membrane is made through the unsaturation of fatty acyl chains, with phosphatidylglycerol (PG) as the main constituent. The presence of high levels of unsaturated FAs in the membranes of the chloroplastic thylakoids can protect the photosynthesis system against chilling-induced photo-inhibition and increases the tolerance of the plant to chilling. Galactolipase, which is responsible for the degradation of membrane lipids, is significantly active in chill-sensitive plants at low temperatures. Common signs of chilling damage in chill-sensitive plants include the bulging of the chloroplast and thylakoid membranes, an increase in the number of plastoglobules (Zbierzak et al., 2013; Karim et al., 2014), darkening of the stroma, unstacking of the grana, vesicle accumulation, and chloroplast disintegration (Karim et al., 2014). In chill-resistant plants, more condensed grana disks are found along with a decrease

14. Responses of plants to cold stress.

Effect of low temperature stress on plants (Primary effects)	Low temperature stress and plant Acclimation	Effect of low temperature stress on plant (Secondary stress)	Synthesis of compounds under acclimatized condition	Plant response
Reduction in water potential Ice crystal formation in cell Cellular dehydration De- stabilization of membrane		Oxidative stress Reduced leaf expansion Stomata closed Photosynthesis inhibition and leaf abscission Reduced cellular and metabolic activity Destabilization of membrane lipids and protenaceous components Ions toxicity Cell death	Accumulation of osmolites: proline/ glycine betain/ trehalose/ increased total amount of soluble sugar Reduction of electrolytic leakage Production of antifreezing protein/ cold shock proteins/ ACC-deaminase enzymes	Induced systematic tolerance in plants by altering host physiology and metabolic responces Increased membrane fluidity Synthesis of antioxidative enzymes and reduced ROS accumulation Ion homeostasis and upregulation of proton ions and calcium pump transporting genes like CsHA2/ CsHA3/CsHA4/CsHA 8/CsHA9/CsHA10/C A1/CA9 etc.

in the number of starch granules over time (Zhuang et al., 2018). The above-mentioned symptoms are closely related to the reprogramming of nuclear and chloroplast-encoded gene expression. Several nuclear genes (like PPR, RNPs, and TCD5) regulate RNA processing at the transcriptional level of the chloroplast and decrease galactolipase activity, which assists the plant in adapting to chilling stress [**13, 14**].

Correlation between heavy metal toxicity and signaling with oxidative stress

Many terrestrial ecosystems have experienced industrialization, with usage of heavy metals that accumulate as residues in the soil and groundwater before being taken up by crop plants, reducing their productivity (Muhammad et al., 2015). Accumulation of heavy metals causes direct and/or indirect reductions in growth due to alteration in the physiological and molecular activities of plants (Hassan et al., 2017). Heavy metals, such as zinc (Zn), copper (Cu), molybdenum (Mo), manganese (Mn), cobalt (Co), and nickel (Ni), are essential to crucial biological processes and developmental pathways. However, these metals, along with four other heavy metals (arsenic (As), lead (Pb), cadmium (Cd), and mercury (Hg)), are considered to be extremely toxic by the Agency for Toxic Substances and Disease Registry (ATSDR, 2003). According to Xiong et al. (2014), these metal ions significantly reduce crop productivity when their concentrations rise above supra-optimal values. These toxic elements cause morphological abnormalities and metabolic disorders that lead to yield reductions in plants (Amari et al., 2017). These abnormalities also give rise to the production of ROS, e.g. superoxide anion radicals (O^{2-}), H_2O_2, and hydroxyl radicals (OH^-), resulting in the disruption of the redox homeostasis of cells (Muhammad et al., 2015) [**15**]. Arsenic (As) is a naturally occurring metal that is found in the environment as a result of volcanic action, the erosion of rocks, and human activities such as the use of pesticides and wood preservatives, and mining and smelting operations. This arsenic seeps through soil and then contaminates groundwater. This contaminated water is then used during crop plant irrigation, subsequently accumulating in different plant tissues, including the grains, before finally entering the food chain (Verma et al., 2016). Inorganic arsenate As (V), arsenite As(III), As(III) and As(V), are the forms of arsenic that exist naturally in the environment. As (III) and As (V) are toxic and are regarded as major environmental pollutants by the United States Environmental Protection Agency (USEPA) (Verma et al., 2017).

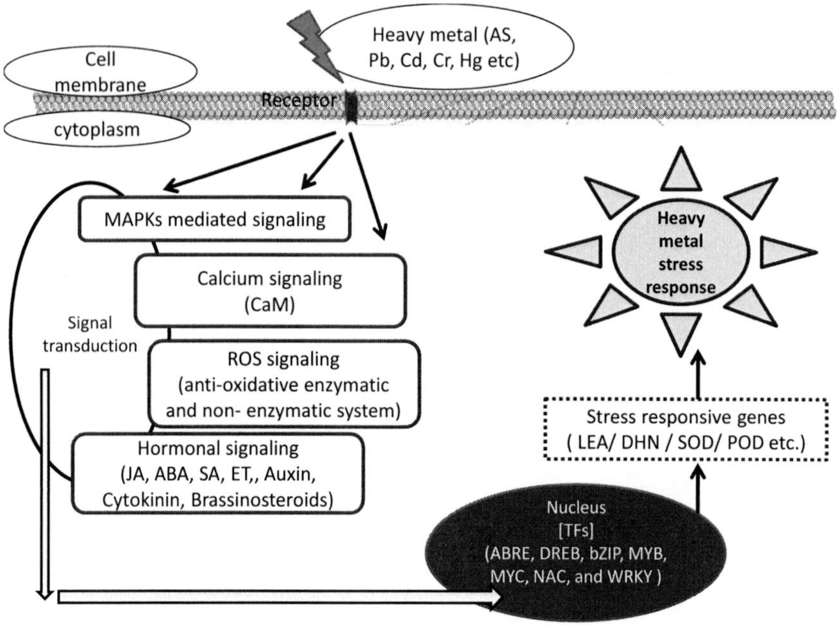

15. Signal transduction in plants under heavy metal stress conditions.

(Elevated levels of metal ions are sensed by a membrane sensor/receptor. The MAPK (mitogen-activated protein kinase) cascades, CaM (calcium signaling pathway), ROS, and hormonal signaling pathways are involved in regulating ionic homeostasis, plant growth and development.)

As (III) is more toxic than As (V) and disrupts biological functions in plants by binding sulfhydryl groups to proteins and interfering with their functions (Verma et al., 2016). It produces ROS, inhibiting the ETC of the respiration process by binding to vicinal thiols in pyruvate dehydrogenase and 2-oxo-glutarate dehydrogenase. In the various metabolic pathways, it indirectly works as a mutagen and causes intra-chromosomal homologous recombination (Helleday et al., 2000). As (V) interferes with oxidative phosphorylation and ATP synthesis during energy metabolism. Lead (Pb^{2+}) is a non-biodegradable metallic form, long-term exposure to which is extremely damaging for the soil and plant growth. High concentrations of Pb ions in plants adversely affect germination, seedling development, root elongation, transpiration, chlorophyll biosynthesis, and cell division (Traunfeld and Clement, 2001; Pourrut et al., 2011).

It causes phytotoxicity and changes cell membrane permeability by reacting with the active groups of different metabolic enzymes and with the phosphate groups of ADP or ATP (Kumar et al., 2017). Pb toxicity leads to inhibition of ATP production and induces lipid peroxidation, resulting in DNA damage due to over-production of ROS.

Cadmium (Cd) is another phytotoxic heavy metal. It is highly soluble in water and so can easily enter the food chain and cause serious human health problems (Buchet et al., 1990). Cd has been classified as a potent human carcinogen by the International Agency for Research on Cancer (IARC, 1993). Low concentrations of Cd can significantly alter the activities of many enzymes involved in the Calvin cycle, carbohydrate and phosphorus metabolism, and CO_2 fixation. This can ultimately result in stunted growth; chlorosis; leaf epinasty; alterations in the chloroplast ultrastructure; inhibition of photosynthesis, pollen germination and tube growth; induction of lipid peroxidation; alterations in nitrogen (N) and sulfur (S) metabolism; and the disruption of antioxidant machinery (Gill and Tuteja, 2011).

Mercury (Hg) is a natural component of the earth's crust. Human activity, such as mining, is responsible for the entry of Hg into the food chain. In the environment, Hg exists in several forms, including elemental (Hg^0) and inorganic (Hg^{2+}), and these forms are associated with other ions (like HgS, $ClHg_2$, Hg_2Cl_2), and organic (CH_3-Hg). High concentrations of Hg are highly phytotoxic to plant cells and can cause visible damage and physiological disorders. In agricultural land, Hg is present in soil, predominantly in the form of Hg^{2+}. The binding of Hg^{2+} to water channel proteins leads to the closure of leaf stomata, resulting in physical obstruction of water flow in plants. Hg^{2+} interferes with mitochondrial activity and induces oxidative stress by stimulating the generation of ROS in plants. This interferes with the bio-membrane lipids and cellular metabolism, as well leading to increased activity of antioxidant enzymes, like SOD, POD, or APX, which can indicate the degree of plant stress (Zhou et al., 2007, 2008).

Heavy metals activate distinct signaling pathways in plants, such as calcium dependent signaling, mitogen-activated protein kinase signaling, ROS signaling, and hormone signaling that enhances the expression of TFs and/or stress-responsive genes (Kumar and Trivedi, 2016). Diverse Ca^{2+} sensors, such as calmodulins (CaMs), CaM-like proteins, calcineurin B-like proteins (CBLs), and Ca^{2+} dependent protein kinases (CDPKs), exist in plants, which sense, decode, and convey alterations in cytosolic

Ca^{2+} concentration for the stress response (Conde et al., 2011) [**15**]. Similarly, the MAPK signaling cascade phosphorylates numerous TFs, such as ABRE, DREB, bZIP, MYB, MYC, NAC, and WRKY, thus influencing metal stress response (Lin and Aarts, 2012; Tiwari et al., 2017a). High levels of Cu and Cd are known to activate distinct MAPKs in *Medicago sativa* (Jonak et al., 2004). ROS are also known to disrupt various phytohormone signaling pathways, including auxin, ethylene, and JA. Variations in the levels of phytohormones change the plant's response during heavy metal stress conditions.

Correlation between ultraviolet-ionizing radiation and oxidative stress

There is ultraviolet (UV) radiation in the environment, which has an adverse impact on plant productivity. Being sessile, plants are continuously exposed to DNA-damaging agents present in the environment, such as those of ultraviolet (UV) and ionizing radiation (IR). Sunlight acts as an energy source for photosynthetic plants; hence UV radiation (namely, UV-A, 315–400 nm; UV-B, 280–315 nm; and UV-C, <280 nm) is unavoidable [**16**]. DNA, in particular, strongly absorbs UV-B; therefore, it is the most important target for UV-B induced damage. On the other hand, IR causes water radiolysis, which generates highly reactive hydroxyl radicals ($OH^{\bullet}$) and causes radiogenic damage to important cellular components. The overproduction of reactive oxygen species (ROS), as byproducts of normal cellular metabolism, leads to DNA damage in the cell through both single-strand DNA breaks (SSBs) and double-strand DNA breaks (DSBs). Intrinsic DNA damage may include the loss of a base to form an abasic site, chemical modification of a base to form a miscoding or noncoding lesion, and sugar-phosphate backbone breakage (Gills and Tuteja, 2010; Biedermann et al., 2011). ROS-altered base and nucleotide sequences have high guanosine content. UV-B radiation penetrates the plant's genome and causes oxidative damage, which results in the formation of pyrimidine hydrates and cross-links between DNA-protein and DNA-DNA. UV-B radiation damages nuclear, chloroplast, and mitochondrial DNA by inducing various DNA lesions, including the generation of the cyclobutane pyrimidine dimers (CPDs) and other photoproducts, pyrimidine (6-4) pyrimidone dimers. The accumulation of this damage (unrecognized and unrepaired DNA damage) may result lethal mutations, which, in turn, reduce plant genome stability, growth, productivity, and threaten the organism's immediate survival. The negative effects of UV-B radiation include deformation of the plant's morphological parameters and thus threaten growth. Exposure to UV-B reduces plant height, leaf area, and plant dry weight; and increases auxiliary branching and leaf curling (Dai et al., 1995; Greenberg et al., 1997; Furness et al., 1999). Lindon (2012) observed that high levels of UV-B and UV-A reduce leaf area per unit of plant biomass, as well as reducing relative growth rate and nitrogen yield. The photosynthetic apparatus of plants can be directly damaged by UV-B. It inactivates the photosystem II (PSII) RUBISCO activase enzyme and sees down-regulation of the transcription of photosynthetic genes. Simultaneously, it decreases the levels of chlorophylls and carotenoids, and impairs the integrity of the thylakoid membrane of the chloroplast. UV-B stress is usually recognized as decreasing photosynthesis and growth, and is associated with alterations in carbon and nitrogen metabolism (Lidon, 2012). It affects stomatal conductance, altering the rate of water loss by transpiration and uptake rate of CO_2 for photosynthesis (Yao and Liu, 2006). This stomatal closure is generated by a loss of turgor pressure with ion leakage from the guard cells. UV-B attenuation is primarily attributed to flavonoids and related phenolic compounds that effectively absorb UV-B radiation while transmitting PAR (photosynthetically active radiation, 400-700 nm) to the chloroplasts (Reuber et al., 1996). The levels of these complex phenolic compounds vary considerably between plant species, according to the plant's developmental stage, and with differing environmental conditions. Exposure to UV-B radiation may increase the concentration of UV-B-absorbing compounds (e.g., flavonoids, carotenoids) in the epidermal cells, rendering some plants less susceptible to photosynthetic damage [**17**]. When plant cells undergo prolonged exposure to UV-B radiation they produce ROS, which induce oxidative damage on DNA. Specific cell signaling pathways and cellular responses are activated, such as the production of stress proteins, upregulation of antioxidants, and accumulation of compatible solutes. In response, and to cope with this oxidative stress, various ROS scavenging systems are found in plants. These are: enzymatic (superoxide dismutase (SOD) catalase (CAT)); and non-enzymatic (ascorbate (ASA), glutathione (GSH), carotenoids (Car), proline and phenols) (Asada, 1999).

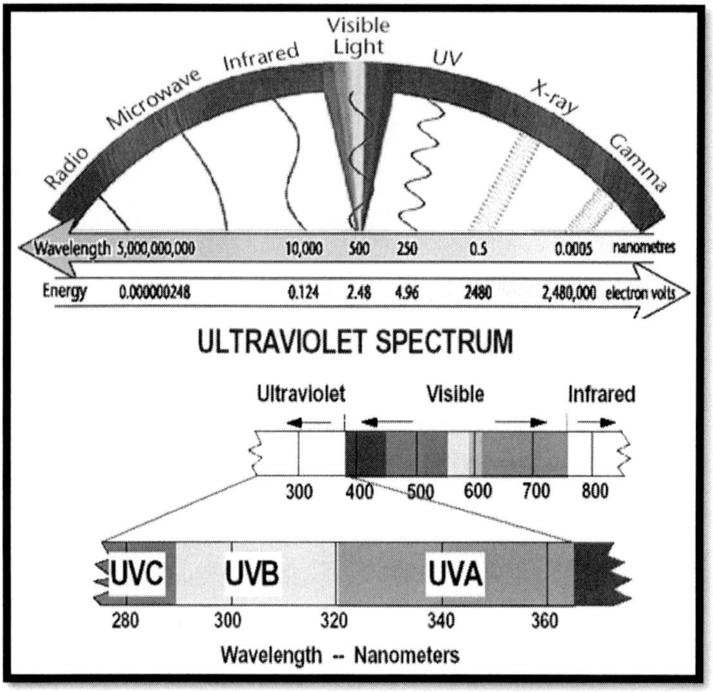

16. The solar spectrum encompasses ultraviolet, visible, and infrared radiation.

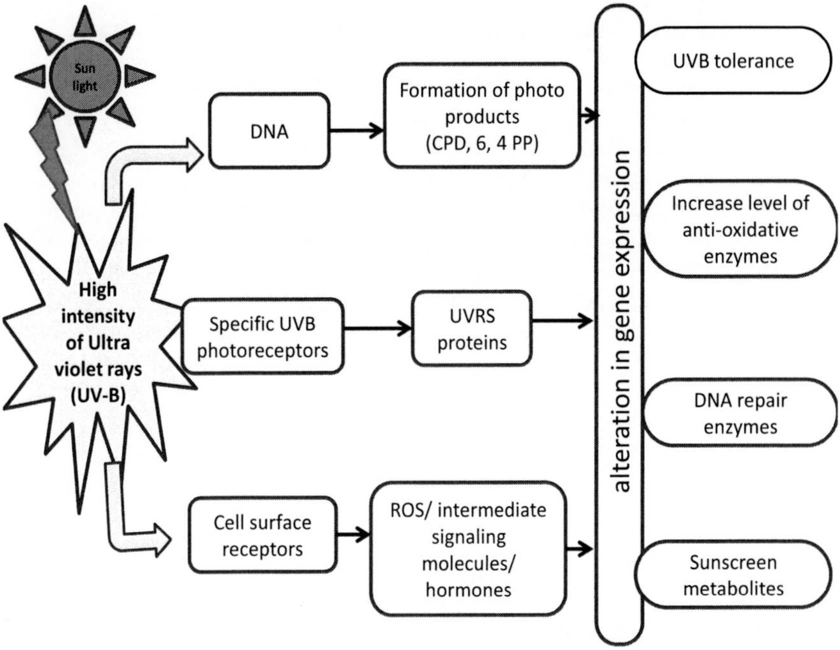

17. Ultraviolet-B radiation mediated responses in plants; balancing damage and protection.

Under conditions of radiation stress, phenolic compounds are metabolically oxidized in the presence of oxidative enzymes, such as POX (EC 1.11.1.7) and PPO (polyphenol oxidase, EC 1.10. 3.1); these catalyze the oxidation of phenols to quinones (Thypyapong et al., 1995). Agarwal (2007) found a decrease in total phenol contents in *Cassia auriculata*, as well as enhanced PPO activity due to UV-B radiation. It is thus clear that an increase in UV-B causes photomorphogenic as well as genetic and physiological changes in plants. Photoreceptors acting through signal transduction pathways are responsible for sensing this ultraviolet radiation. Several components of the photosynthetic apparatus have been found to be affected by UV-B, with nuclear-encoded genes being more sensitive to UVB than chloroplast-encoded genes.

Apoplastic oxidative bursts in response to biotic stress

Plants are continuously exposed to a variety of microorganisms that are present in the soil. In order to successfully avoid infection, they have evolved a series of defense mechanisms that work in concert to limit pathogen invasion and multiplication. Unlike vertebrates, plants lack an adaptive immune system and rely on their innate immune system to recognize and restrict pathogenic microbes. Conceptually, there are two primary branches of innate plant immunity. One branch employs extracellular receptors to recognize conserved microbial features, termed pathogen-associated molecular patterns (PAMPS), resulting in PAMP-triggered immunity (PTI). The second branch uses intracellular plant resistance (R) proteins to recognize pathogen effectors inside host cells during infection, resulting in effector-triggered Immunity (ETI). The activation of PTI leads to the induction of mitogen-activated protein kinase (MAPK) signaling, transcriptional reprogramming, ROS production, and callose deposition, which serves to create a physical barrier at the infection site [**18**]. In the initial cellular response, an ROS or oxidative burst is rapidly produced in the plant followed by successful pathogen recognition. Radwan et al. (2010) observed higher H_2O_2 and MDA concentrations in *Vicia faba* leaves infected with bean yellow mosaic virus than those of corresponding control plants. Several enzymes have been implicated in apoplastic ROS production followed by successful pathogen recognition. Plasma membrane NADPH oxidases and cell wall peroxidases are enzymes whose expression levels are used to recognize bacterial and fungal infections in plant cells. Although the primary oxidative burst of pathogen recognition occurs in the apoplast, the mitochondria and chloroplast are cellular compartments that also produce ROS under biotic stress. Liu et al. (2007) showed that activation of the SIPK/NTF4/WIPK cascade by pathogens actively promotes the generation of ROS in the chloroplast, which plays an important role in plant signal generation. They concluded that chloroplast bursts occur earlier than NADPH oxidase bursts and that mitochondria-generated ROS may be vital to the acceleration of cell death. In tobacco, the reduction of CAT and APX activity results in the plants becoming hyper responsive to pathogens. A significant increase in the activity of POD and CAT was observed in the leaves of flax lines infected with powdery mildew (Ashry and Mohamed, 2012). Enhanced POD, CAT, APX, and SOD activity were observed in *Vicia faba* leaves infected with bean yellow mosaic virus indicating that ROS scavenging systems can have an important role in managing ROS generation in response to pathogens (Radwan et al., 2010).

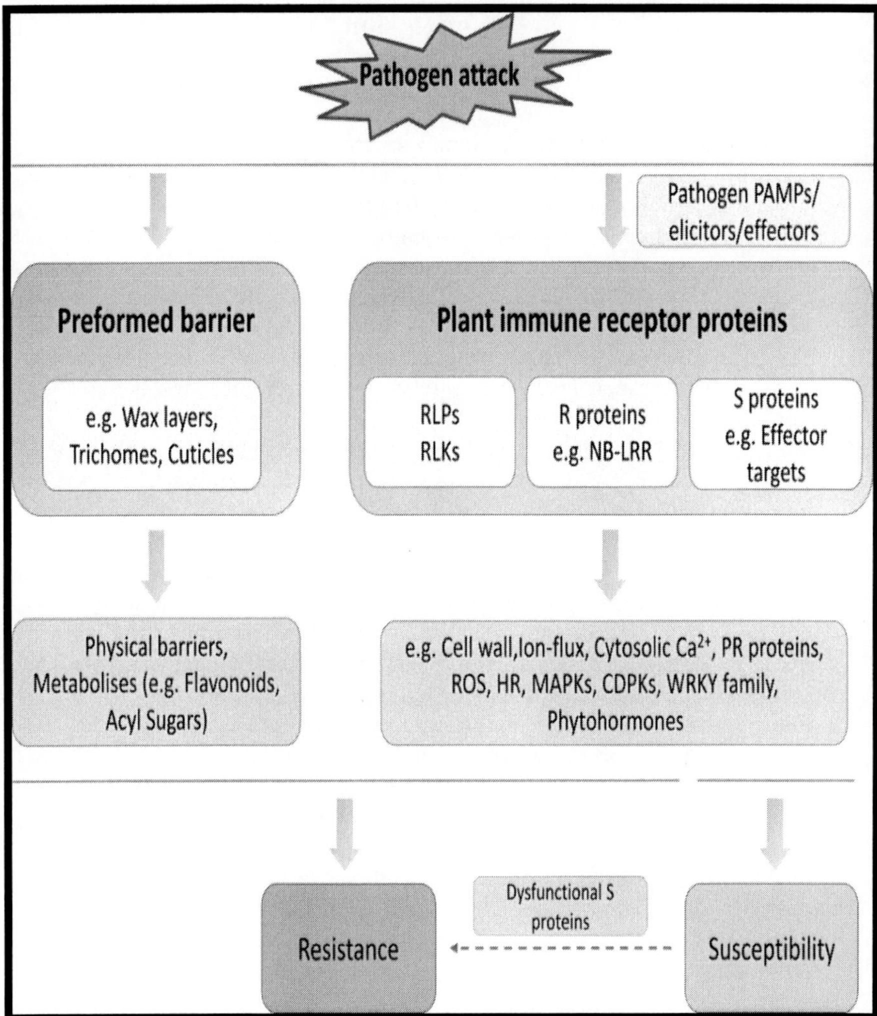

18. Outline of strategies for improving crop performance under pathogenic stress.

Conclusion

ROS are unavoidable byproducts of normal cell metabolism. ROS are generated either by the electron transport activity of the chloroplast, mitochondria, and plasma membrane or byproducts of various metabolic pathways that are localized in the different cellular compartments. During normal growth conditions, there is low production of ROS in the various cell compartments. However, when plants are exposed to various environmental stresses, such as those of drought, salinity, chilling, heavy metal toxicity, and UV-B radiation, disruption of cellular homeostasis due to increased ROS production results. ROS play two divergent roles in plants: in low concentrations, they act as signaling molecules that mediate several plant responses in plant cells, including responses to stresses; however, in higher concentrations, they cause damage to the cellular components. High levels of ROS cause oxidative damage to plant lipids, proteins, and DNA leading to altered intrinsic membrane properties concerning fluidity, ion transport, loss of enzyme activity, protein cross-linking, inhibition of protein synthesis, and DNA damage, ultimately causing cell death. In order to avoid this oxidative damage, higher plants possess a complex antioxidative defense system comprising nonenzymatic and enzymatic components. There remain many uncertainties and gaps in our knowledge about ROS formation and their effect on plants, mainly due to their short half-life and high reactivity. Future progress in genomics, metabolomics, and proteomics should increase our understanding of the biochemical networks involved in cellular responses to oxidative stress. There is a need to study the formation and fate of ROS using advanced analytical techniques, to enable mapping of the functionality of ROS for the sustenance and growth of plants. Biotechnological techniques will enable a better understanding of the patterns of behavior of ROS during various stress conditions, inducing enzymatic and non-enzymatic antioxidative actions that can help plants and crops to grow in stressful environmental conditions.

References

Agarwal, Shivansh. 2007. "Increased antioxidant activity in *Cassia* seedlings under UV-B radiation". *Biologia Plantarum*, No. 51(1): 157-160.

Agency for Toxic Substances and Disease Registry (ATSDR) (2003). http://www.atsdr.cdc.gov//.

Ali, Shaik, Sandhya, Vardharajula, and Rao, Linga. 2014. "Isolation and characterization of drought-tolerant ACC deaminase and exo-polysaccharide-producing fluorescent *Pseudomonas sp*". *Annals of Microbiology*, No. 64: 493-502.

Amari, Taoufik, Tahar, Ghnaya and Chedly, Abdel. 2017. "Nickel, cadmium and lead phytotoxicity and potential of halophytic plants in heavy metal extraction". *South African Journal of Botany*, No. 111: 99-110.

Apel, Klaus and Hirt, Heribert. 2004. "Reactive oxygen species: metabolism, oxidative stress, and signal transduction". *Annual Review of Plant Biology*, No. 55: 373-399.

Arora, Ajay, Sairam, Raj Kumar and Srivastava. 2002. "Oxidative stress and antioxidative system in plants". *Current Science*, No. 82(10): 1227-1238.

Asada, Kozi. 1999. "The water–water cycle in chloroplasts: scavenging of active oxygen and dissipation of excess photons". *Annual Review of Plant Physiology*, No. 50: 601-639.

Ashry, Naglaa and Mohamed, Heba. 2012. "Impact of secondary metabolites and related enzymes in flax resistance and/or susceptibility to powdery mildew". *African Journal of Biotechnology*, No. 11(5): 1073-1077.

Astorga Gloria, and Melendez, Lilia. 2010. "Salinity effects on protein content, lipid peroxidation, pigments and proline in *Paulownia imperialis* and *Paulowina fortune* grown *in vitro*". *Electronic Journal of Biotechnology*, No. 13(5): 13-14.

Baxter, Brugha, Erskine, Scheurer, Vos and Scott. 2014. "The epidemiology and global burden of autism spectrum disorders". *Psychological medicine*, No. 45(3): 601-13.

Beinsan, Carmen, Dorin, Camen, Radu, Sumalan and Mariana, Babau. 2003. "Study concerning salt stress effect on leaf area dynamics and chlorophyll content in four bean local land races from Banat areas". *Faculty of Horticulture*, No. 119: 416-419.

Bernstein, Nirit, Meiri and Zilberstaine, Miriam. 2004. "Root growth of avocado is more sensitive to salinity than shoot growth". *Journal of the American Society for Horticultural Science,* No. 129: 188-192.

Biehler and Fock. 1996. "Evidence for the contribution of the Mehler-peroxidase reaction in dissipating excess electrons in drought-stressed wheat", *Plant Physiology*, No. 112(1): 265-272.

Blokhina, Olga and Fagerstedt, Kurt. 2010. "Reactive oxygen species and nitric oxide in plant mitochondria: origin and redundant regulatory systems". *Physiologia Plantarum*, No. 138(4): 447-462.

Buchet, Jean Pierre, Lauwerys, Robert, Roels, Harry, Bernard, Alferd, Bruaux, Pierre and Claeys, Francoise. 1990. "Renal effects of cadmium body burden of the general population". *The Lancet*, No. 336 (8717): 699-702.

Choluj, Danuta, Karwowska, Jasinska and Haber. 2004. "Growth and dry matter partitioning in sugar beet plants (*Beta vulgaris* L.) under moderate drought". *Journal of Plant Soil and Environment*, No. 50 (6): 265-272.

Conde, Artur, Silva, Paulo, Agasse, Alice, Conde, Carlos and Geros, Hernani. 2011. "Mannitol transport and mannitol dehydrogenase activities are coordinated in Olea europaea under salt and osmotic stresses". *Plant Cell Physiolology*, No. 52(10): 1766-75.

Cosio, Claudia and Dunand, Christophe. 2009. "Specific functions of individual class III peroxidase genes". *Journal of Experimantal Botany*, No. 60(2): 391-408.

Couee, Ivan, Sulmon, Cecile, Gouesbet, Rennes and Amrani. 2006. "Involvement of soluble sugars in reactive oxygen species balance and responses to oxidative stress in plants". *Journal of Experimental Botany*, No. 57(3): 449-59.

Dai, Aiguo, Trenberth, Kevin and Karl Thomas. 1998. "Global Variations in Droughts and Wet Spells: 1900-1995". *Geophysical Research Letters*, No. 17 (25): 3367-3370.

Dai, Qiujie, Shaobing, Peng, Arlene, Chavez and Benito, Vergara. 1995. "Effects of UV-B radiation on stomatal density and opening in rice (*Oryza sativa* L.)". *Annals of Botany*, No. 76: 65-70.

Del, Rio, Sandalio, Luisa, Corpas, Francisco, Palma, Jose and Barroso, Juan. 2006. "Reactive oxygen species and reactive nitrogen species in peroxisomes production, scavenging, and role in cell signaling". *Plant Physiology*, No. 141(2): 330-5.

Dietz, Megan, Teresa, Bernaciak, Vendetti, Frank, Kielec, Joseph and Hildebrand, Jeffrey. 2006. "Differential actin-dependent localization modulates the evolutionarily conserved activity of Shroom family proteins". *Journal of Biological Chem*istry, No. 281: 20542-20554.

Foyer, Christine. 1997. "Oxygen metabolism and electron transport in photosynthesis". *In: Molecular Biology of Free Radical Scavenging Systems*. Journal of Scandalios. 587-621. Cold Spring Harbor Laboratory Press, (New York, NY, USA).

Furness, Nancy, Upadhyaya, Mahesh and Ormrod Douglas. 1999. "Seedling growth and leaf surface morphological responses of three rangeland weeds to ultraviolet-B radiation". *Weed Science*, No. 47: 427-434.

Gill, Sarvajeet and Tuteja, Narendra. 2011. "Cadmium stress tolerance in crop plants: probing the role of sulfur". *Plant Signaling & Behaviour*, No. 6 (2): 215-222.

Gill, Sarvajeet and Tuteja, Narendra. 2010. "Reactive oxygen species and antioxidant machinery in abiotic stress tolerance in crop plants". *Plant Physiology and Biochemistry*, No. 48(12): 909- 30.

Greenberg BM, Wilson MI, Huang XD, Duxbury CL, Gerhaddt KE and Gensemer RW (1997). "The effects of ultraviolet-B radiation on higher plants". In: Wang W, Goursuch J and Hughes JS (eds.): *Plants for environmental studies*. Boca Raton, Fl: CRC Press: 1-35.

Gross, Georg. 1977. "Cell wall-bound malate dehydrogenase from horseradish". *Phytochemistry*, No. 16(3): 319-321.

Han, Chao, Liu, Qing and Yang, Yan. 2009. "Short-term effects of experimental warming and enhanced ultraviolet-B radiation on photosynthesis and antioxidant defense of *Picea asperata* seedlings". *Plant Growth Regulation*, No. 58(2): 153-162.

Hassan, Tamoor, Bano, Asghari and Naz, Irum. 2017. "Alleviation of heavy metals toxicity by the application of plant growth promoting rhizobacteria and effects on wheat grown in saline sodic field". *Internationational Journal of Phytoremediation*, No. 19(6): 522-529.

Heidari, Mostafa and Golpayegani, Amir. 2011. "Effects of water stress and inoculation with plant growth promoting rhizobacteria (PGPR) on antioxidant status and photosynthetic pigments in basil (*Ocimum basilicum* L.)". *Journal of the Saudi Society of Agricultural Sciences*, No. 11: 57-61.

Helleday, Thomas, Nilsson, Robert and Jenssen, Dag. 2000. "Arsenic (III) and heavy metal ions induce intra chromosomal homologous recombination in the *hprt* gene of V79 Chinese hamster cells". *Environmental and Molecular Mutagenesis*, No. 35: 114-122.

Heyno, Eiri, Veronique, Mary, Schopfer, Peter and Krieger-Liszkay, Anja. 2011. "Oxygen activation at the plasma membrane: relation between superoxide and hydroxyl radical production by isolated membranes". *Planta*, No. 234 (1): 35-45.

IARC. 1993. "IARC Monographs on the Evaluation of Carcinogenic Risk to Humans". *Beryllium, Cadmium, Mercury and Exposures in the Glass Manufacturing Industry*. 58: 41-117. Lyon: International Agency for Research on Cancer Publications.

Jonak, Claudia, Nakagami, Hirofumi and Hirt, Heribert. 2004. "Heavy metal stress. Activation of distinct mitogen-activated protein kinase pathways by copper and cadmium". *Plant Physiology*, No. 136(2): 3276-3283.

Karim, Sazzad, Mohamed, Alezzawi, Christel, Garcia-Petit, Katalin, Solymosi, Khan, Nadir, Emelie, Lindquist, Dahl, Peter, Stefan, Hohmann and Henrik Aronsson. 2014. "A novel chloroplast localized Rab GTPase protein CPRabA5e is involved in stress, development, thylakoid biogenesis and vesicle transport in Arabidopsis". *Plant Molecular Biology*, No. 84: 675-692.

Keunen, Els, Peshev, Darin, Vangronsveld, Jaco, Ende, Wim and Cuypers, Ann. 2013. "Plant sugars are crucial players in the oxidative challenge during abiotic stress: extending the traditional concept". *Plant Cell and Environment*, No. 36(7): 1242-1255.

Kumar Smita and Trivedi Prabodh.2016. "Heavy Metal Stress Signaling in Plants". In: *Plant Metal Interaction*, 585-603.

Kumar, Brajesh, Kumari, Smita and Flores, Luis. 2017. "Plant mediated detoxification of mercury and lead". *Arabian Journal of Chemistry*, No. 10(2): S2335-S2342.

Lidon, Fernando. 2012. "Micronutrients' accumulation in rice after supplemental UV-B irradiation". *Journal of Plant Interactions*, 7(1): 19-28.

Lin, Fen and Aarts, Mark. 2012. "The molecular mechanism of zinc and cadmium stress response in plants". *Cellular and Molecular Life Sciences*, No. 69(19): 3187-206.

Liu, Li, Yang, Sui, Yang, and Meng. 2008. "Over expression of tomato chloroplast omega-3 fatty acid desaturase gene alleviates the photo-inhibition of photo-systems 2 and 1 under chilling stress". *Photosynthetica*, No. 46 (2): 185-192.

Liu, Yidong, Dongtao Ren, Sharon Pike, Stephen Pallardy, Gassmann, Walter and Zhang, Shuqun. 2007. "Chloroplast-generated reactive oxygen species are involved in hypersensitive response-like cell death mediated by a mitogen-activated protein kinase cascade". *The Plant Journal*, No. 51(6): 941-954.

Maheshwari, Ruchi, Dubey, Shanker. 2009. "Nickel-induced oxidative stress and the role of antioxidant defence in rice seedlings". *Plant Growth Regulation*, No. 59(1): 37-49.

Martin, Weih, Johanson, Ulf and Dylan, Jones. 1998. "Growth and nitrogen utilization in seedlings of mountain birch (*Betula pubescens* ssp. tortuosa) as affected by ultraviolet radiation (UV-A and UV-B) under laboratory and outdoor conditions". *Trees*, No. 12: 201-207.

Martinez, Montillet, Bresson, Agnel, Dai, Daniel, Geiger and Nicole. 1998. "Apoplastic peroxidase generates superoxide anions in cells of cotton cotyledons undergoing the hypersensitive reaction to

Xanthomonas campestris pv. *malvacearum* 18." *Plant-Microbe Interactions*, No. 11(11): 1038-1047.

Maxwell, Patrick, Michael, Gin-Wen, Steven, Emma, Cockman, Matthew, Wykoff, Charles, Pugh, Christopher, Maher, Eamonn and Ratcliffe, Peter. "The tumour suppressor protein VHL targets hypoxia-inducible factors for oxygen-dependent proteolysis". *Nature*, No. 399(6733): 271-275.

Meyer, Kate, Saletore, Yogesh, Zumbo, Paul, Elemento, Olivie, Mason, Christopher and Jaffrey, Samie. 2012. "Comprehensive analysis of mRNA methylation reveals enrichment in 3' UTRs and near stop codons". *Cell*, No. 149(7): 1635-46.

Mittler, Ron. 2002. "Oxidative stress, antioxidants and stress tolerance". *Trends in Plant Science*, No. 7(9): 405-410.

Muhammad, Shahid, Khalid, Sana, Abbas, Ghulam, Naeem, Shahid, Muhammad, Nadeem, Muhammad, Sabir, Dr-Muhammad, Aslam and Dumat, Camille. 2015. "Heavy metal stress and crop productivity". In *Crop Production and the Global Environment*, edited by Rehman Hakeem Khalid, 1: 1-25. Springer International Publishing.

Nair, Akhila, Abraham and Jaya. 2008. "Studies on the changes in lipid peroxidation and antioxidants in drought stress induced Cowpea (*Vigna unguiculata* L.) varieties". *Journal of Environmental Biology*, No. 29: 689-691.

Noctor Graham, Trouverie, Caroline and Mhamdi, Amna. 2014. "The metabolomics of oxidative stress". *Phytochemistry*, No. 112(1): 33-53.

Pourrut, Bertrand, Severine, Jean, Jerome, Silvestre and Eric, Pinelli. 2011. "Lead-induced DNA damage in *Vicia faba* root cells: potential involvement of oxidative stress". *Mutagenesis*, No. 726(2): 123-128.

Radwan, Mohammed, Deya, Eldeen, Fayez, Khalaf Ali, Mahmoud, Sabry Younis and Lu, Guoquan. 2010. "Modifications of antioxidant activity and protein composition of bean leaf due to *Bean yellow mosaic virus* infection and salicylic acid treatments". *Acta Physiologiae Plantarum*, No. 32(5): 891-904.

Rahdari, Parvaneh, Hosseini, Seyed and Tavakoli Shahrokh. 2012. "The studying effect of drought stress on germination, proline, sugar, lipid, protein and chlorophyll content in Purslane (*Portulaca oleraceae* L.) leaves". *Journal of Medicinal Plant Research*, No. 6 (9): 1539-1547.

Reuber, Sebastian, Bornman, Janet and Weissenbock, Gottfried. 1996. "A flavonoid mutant of barley (*Hordeum vulgare* L.) exhibits increased sensitivity to UV-B radiation in the primary leaf". *Plant Cell and Environment*, No. 19(5): 593-601.

Sascha, Mooney, Sutton and Hellmann Hanjo. 2011. "Recognition and repair pathways of damaged DNA in higher plants". *DNA Repair*, Sascha Biedermann, 201-236. (USA: Washington State University).

Sgherri, Cristina, Cecconami, Silvia, Pinzino, Calogero, Izzo, avia and Izzo, Riccardo. 2010. "Antioxidant activity and nutraceuticals in basil grown in hydroponics and soil". *Food Chemistry*, No. 123(2): 416-422.

Shah, Kavita, Kumar, Ritambhara, Verma, Shalini and Dubey. 2001. "Effect of cadmium on lipid peroxidation, superoxide anion generation and activities of antioxidant enzymes in growing rice seedlings". *Plant Science*, No. 161(6): 1135-1144.

Sharma, Pallavi and Dubey, Shanker. 2005. "Drought induces oxidative stress and enhances the activities of antioxidant enzymes in growing rice seedlings". *Plant Growth Regulation*, No. 46(3): 209-221.

Siddiqi, Ejaz, Ashraf, Muhammad, Hussain, Mumtaz and Jamil, Amer. 2009. "Assessment of inter cultivar variation for salt tolerance in safflower (*Carthamus tinctorius* L.) using gas exchange characteristics as selection criteria". *Pakistan Journal of Botany*, No. 41(5): 2251-2259.

Tanou, Georgia, Molassiotis, Athanassios and Diamantidis, Grigorios. 2009. "Induction of reactive oxygen species and necrotic death-like destruction in strawberry leaves by salinity". *Environmental and Experimental Botany*, No. 65(2-3): 270-281.

Thipyapong, Piyada, Hunt, Michelle and Steffens, John. 1995. "Systemic wound induction of potato (*Solanum tuberosum*) polyphenol oxidase". *Phytochemistry*, No. 40(3): 673-676.

Tiwari, Meenakshi and Chaube, Shail. 2017a. "Maturation promoting factor destabilization mediates human chorionic gonadotropin induced meiotic resumption in rat oocytes". *Development, Growth and Differentiation*, No. 59(7): 603-614.

Traunfeld, Jon and Clement. 2001. "Lead in Garden Soils". *Home and Garden*. College Park, MD: University of Maryland. Maryland Cooperative Extention.

Verma, Pankaj, Verma, Shikha, Pande, Veena, Mallick, Shekhar, Tripathi, Rudra, Deo, Dhankher, Om and Chakrabarty, Debasis. 2016. "Over expression of rice gluta-redox in OsGrx C7 and OsGrx C2 reduces intracellular arsenic accumulation and increases tolerance in *Arabidopsis thaliana*". *Frontier in Plant Science*, No. 7: 740.

Verma, Shikha, Varma, Pankaj, Varma, Alok, Bansiwal, Amit, Tripathi, Rudra Deo and Chakrabarty, Debasis. 2017. "A novel fungal arsenic methyl transferase, Waars M reduces grain arsenic accumulation in the transgenic rice plant". *Journal of Hazardous Materials*, No. 344: 626-634.

Verma, Shikha, Verma, Pankaj, Meher, Alok, Dwivedi, Sanjay, Bansiwal, Amit and Pande, Veena. 2016. "A novel arsenic methyl transferase gene of *Westerdykella aurantiaca* isolated from arsenic contaminated soil: phylogenetic, physiological, and biochemical studies and its role in arsenic bioremediation". *Metallomics*, No. 8(3): 344-353.

Wang, Shuai, Bai, Ge, Wang, Shu, Yang, Leiyun, Yang, Fen, Wang, Yi, Zhu, Jian and Hua, Jian. 2016. "Chloroplast RNA-binding protein RBD1 promotes chilling tolerance through 23S rRNA processing in Arabidopsis". *PLoS Genetics*, No. 2(5): e1006027.

Xiong, Tiantian, Leveque, Thibaut, Muhammad, Shahid, Yann, Foucault, and Mombo, Dumat. 2014. "Lead and cadmium phyto availability and human bio accessibility for vegetables exposed to soil or atmospheric pollution by process ultrafine particles". *Journal of Environmental Quality*, No. 43(5): 1593-1600.

Yao, Xiaoqin and Liu, Qing. 2006. "Changes in morphological, photosynthetic and physiological responses of Mono Maple seedlings to enhanced UV-B and to nitrogen addition". *Plant Growth Regulation*, No. 50: 165-177.

Zbierzak, Maria, Porfirova, Svetlana, Griebel, Thomas, Melzer, Michael, Parker, Jane and Dormann, Peter. 2013. "A TIR–NBS protein encoded by *Arabidopsis Chilling Sensitive* 1 (*CHS1*) limits chloroplast damage and cell death at low temperature". *In Plant Journal*, No. 75: 539-552.

Zhou, Sheng, Huang, Si, Guo, Kai, Mehta, Surya, Zhang, Peng and Yang, Zhi. 2007. "Metabolic adaptations to mercury-induced oxidative stress in roots of *Medicago sativa* L". *Journal of Inorganic Biochemistry*, No. 101(1): 1-9.

Zhou, Sheng, Wang, Shao and Yang Zhi. 2008." Biological detection and analysis of mercury toxicity to alfalfa (*Medicago sativa*) plants". *Chemosphere*, No. 70(8): 1500-1509.

Zhuang, Kunyang, Kong, Fanying, Zhang, Song, Meng, Chen, Yang, Minmin, Liu, Zhuangbin, Wang, Yong, Ma, Nana, and Meng, Qingwei. 2018. "Whirly1 enhances tolerance to chilling stress in tomato via protection of photosystem II and regulation of starch degradation". *New Phytolologist*, No. 221(4): 1998-2012.

CHAPTER THREE

THE EFFECTS OF ABIOTIC STRESS ON PLANTS

S. DUHAN

Stress refers to any factor of the environmental conditions that adversely affects plant growth. Stress may be due to the presence of living organisms, such as fungi, bacteria, nematodes, and herbivores, for which the term biotic stress is used. Other reasons for stress can include nonliving factors, such as environmental conditions like drought, waterlogging, cold, heat, hypersalinity, and heavy metals, which are known as abiotic stresses. Abiotic stress is considered more harmful, as environmental conditions are unavoidable for plants due to their sessile nature. Across the world, significant losses in crop yields are recorded due to abiotic stresses. When plants come into contact with any type of abiotic stress, wilting and yellowing of leaves are initial symptoms that can worsen if ignored. Various types of cellular changes, such as chlorophyll degradation, excessive loss of water, hindrance of water uptake, cell membrane disintegration, peroxidation of lipids, proteins, and nucleic acids, and the formation of reactive oxygen species (ROS), i.e., superoxide radicals (O_2^-), hydroxyl radicals (OH^-), and hydrogen peroxide (H_2O_2), occur in plants as a result of abiotic stresses, further causing oxidative stress in plants. Plants have developed several strategies to cope with stress, such as the production of antioxidative enzymes, desaturation of membrane lipids, induction of molecular chaperones, and accumulation of compatible solutes. In this chapter, the changes that occur in plants when they come into contact with various abiotic stresses are discussed.

Introduction

Stress is any unfavourable condition that limits a plant's growth, yield, or productivity (He et al., 2018). Stress is basically of two types: biotic and abiotic. Biotic stress involves living organisms, such as bacteria, fungi, and herbivores, in creating the stress condition. Stress conditions created due to nonliving factors, such as drought, waterlogging, cold, heat, hypersalinity, and the presence of heavy metals, are abiotic (Gull et al., 2019). Stress has harmful effects on both animals and plants, but the inability to move makes plants more susceptible to stress. When plants come into contact with a stressor, they can undergo a wide range of responses, like altered genes, decreased growth and productivity, and death.

Water stress, mainly in terms of drought and salinity, has been identified as the most common stressor limiting plant growth (Araus et al., 2002). Drought and salinity are the primary causes of the increasing salinization of arable lands (Wang et al., 2001). The increasing salinization of arable land is expected to make 30 % of land barren by 2040, reaching 50 % by 2050 (Wang et al. 2003). When a plant comes into contact with a stressor, it undergoes various types of cellular changes, such as wilting, chlorophyll degradation, lipid peroxidation, and membrane disintegration, which ultimately leads to death. Other biochemical changes, like the formation of ROS, i.e. superoxide radicals (O_2^-), hydroxyl radicals (OH^-), and H_2O_2, also takes place, which further causes oxidative stress in plants. Reactive oxygen species cause cellular damage when produced in excessive amounts and can result in death. Plant responses to stress also depend on the timing of the stressor. Plants exposed to stress at the reproductive stage show greater damage than plants exposed to stress at the vegetative stage. Duhan et al. (2018) performed an experiment with four pigeon pea genotypes (ICPH-2431, PARAS, UPAS-120, H09-33). The plants were subjected to waterlogging, salinity, and combined stress at twenty and forty days after sowing. Parameters, including percentage of survival, chlorophyll content, membrane injury, lipid peroxidation, RWC, proline content, and yield attributes, were recorded. A greater decline in the percentage of survival, chlorophyll content, RWC, and yield was seen when the plants were subjected to stress forty days after sowing. Membrane injury was also greater in forty day old plants compared to twenty day old plants. Twenty day old plants showed more tolerance compared to forty day old plants with high proline content. Depending on their response to stress, plants can be categorized into two types: sensitive and tolerant. Sensitive plants are those that show greater damage when exposed to stress and struggle to survive. On the other hand, tolerant plants show tolerance of stress and are better able to survive. Various studies have been conducted with both tolerant and sensitive plants and have found that tolerant plants have better survival abilities even under conditions of severe stress. Duhan et al. (2017a) performed an experiment with four pigeon pea genotypes (ICPH-2431, PARAS, UPAS-120, HO

933) that were raised in polythene bags filled with half a kilogram of a soil+FYM manure mixture. Waterlogging, salinity, and waterlogging+salinity treatments were given for twelve days and observations were recorded at one and eight days after removal from treatment in twenty and forty day old plant leaves. Tolerant genotypes (ICPH-2431, PARAS) showed better survival with a higher increase in antioxidative enzymes and metabolites when compared to sensitive genotypes (UPAS-120, HO 933) under the stress conditions. Plants tend to survive under these unfavourable conditions by developing tolerance to stress. Generalized defense mechanisms against stress include the formation of a thick cuticle, unsaturated fatty acids, reactive species scavengers, molecular chaperones, and compatible solutes inside cells. In addition, low molecular weight compounds, like salicylic acid, jasmonic acids, ethylene, and abscisic acid, play a pivotol role in developing stress tolerance. Beside these, some anatomical changes, like the formation of aerenchyma, also play an important role in enhancing a plant's tolerance of stress. Today, abiotic stresses have become a major threat to crop yields and productivity. This chapter discusses the effect of abiotic stresses on plants.

Effect of abiotic stresses on plants

Abiotic stresses are of various types and are generally due to environmental factors like water, temperature, heavy metals, and salinity, etc. Depending on the factor responsible for stress, the response of the plant varies. Some common responses shown by plants include a lack of seed germination, growth retardation, wilting, senescence, and death [**19**].

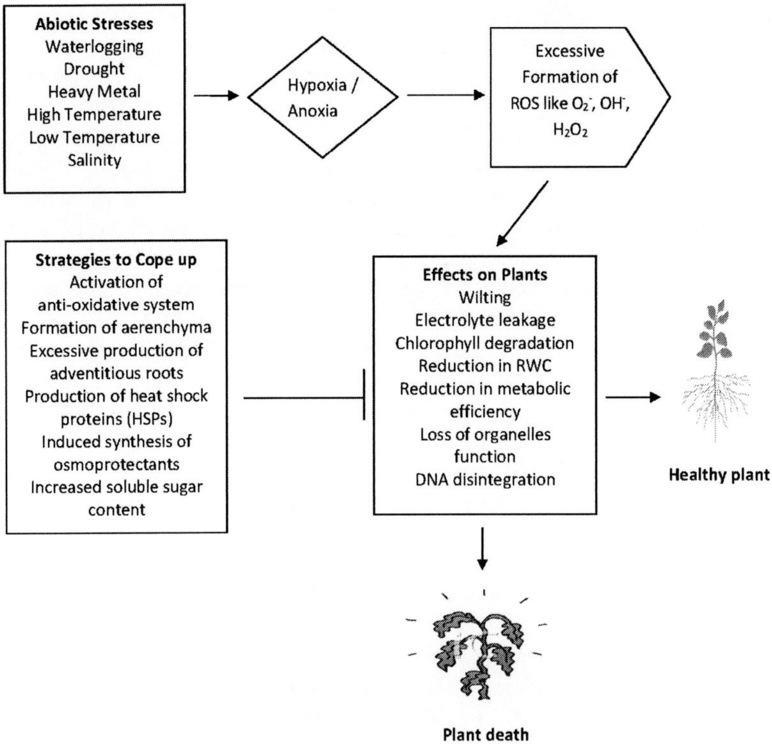

19. Effect of abiotic stresses on plants and their strategies to cope.

Waterlogging

Waterlogging is a serious abiotic stress that occurs due to poor soil drainage and/or excessive rainfall. A vast area of the world is affected by waterlogging—about 10 % of the global land area (Setter and Waters, 2003). Waterlogging reduces crop yields by as much as 80 % (Shabala, 2011). Waterlogged plants face a number of problems, such as limitation of gas exchange, mineral nutrient deficiencies, oxidative stress, and impaired carbohydrate metabolism (Setter et al., 2009; Duhan et al., 2016; Duhan et al 2019; Duhan et al., 2020). Disadvantageous physiological changes, such as decreased survival, chlorophyll content, plant biomass, seed yield per plant, and test weight, have been observed in pigeon pea plants under waterlogging stress conditions (Duhan et al., 2018). Kumutha et al. (2009) observed changes, such as the yellowing and senescence of leaves, membrane instability, and decreased leaf area, dry matter, RWC, and chlorophyll content, in pigeon pea plants. Changes, like decreased root growth, reduced dry matter accumulation, prematurely senescing leaves, reduced tillering, wilting, the production of sterile florets, and low kernel weights and grain yield, have been observed in waterlogged wheat plants (Zhang et al., 2006; Jiang et al., 2008 Hossain et al., 2011). A major consequence of waterlogging is the closure of stomata in plants, which decreases the exchange of gases and also affects passive absorption of water (Aldana et al., 2014). The closure of stomata decreases transpiration, subsequently reducing water absorption, and causes wilting and early senescence of leaves and, ultimately, leaf fall (Kozlowski and Pallardy, 1997). Sairam et al. (2008) found that waterlogging blocks the uptake of nutrients and water by creating an oxygen-deficient environment around the roots so that plants wilt even in the presence of plenty of water. The underlying reason for waterlogging stress are conditions of hypoxia and anoxia, which are created due to presence of water around the root interfering with root respiration. The condition of oxygen unavailability produces free radicals like superoxide radicals (O_2^-), hydroxyl radicals (OH^-), and H_2O_2, further creating oxidative stress if they are not successfully scavenged (Mittler et al. 2004; Tripathy and Oelmuller 2012; Mittler 2017). Another reason for ROS production is the reaction of oxygen molecules with chlorophyll molecules, which produces singlet oxygen (Ashraf and Akram, 2009). These free radicals are highly reactive and cause peroxidation of the lipids and proteins of cell membranes, resulting in membrane disruption (Ashraf, 2012). This exposes nucleic acid to ROS, which is then also peroxidised. ROS are also produced under normal conditions, but production becomes much greater under stress conditions. This is detrimental to cellular metabolic reactions like photosynthesis, affecting PS II (Ashraf, 2009) and the Calvin cycle (Yadav and Atri, 2017). Plants have developed some effective mechanisms for survival against the adverse impacts of ROS. One of these is the production of two types of antioxidants: enzymatic and non-enzymatic. APX, SOD, POD, CAT, and GR are all enzymatic antioxidants, whereas ascorbic acid, glutathione, tocopherols, and carotenoids are all non-enzymatic antioxidants (Gupta et al., 2005). Alterations have been noticed in many plants through the activity of different enzymatic and non-enzymatic antioxidants during waterlogging. Pigeon pea plants showed increased activity of various enzymatic antioxidants, such as GR, SOD, CAT, and APX, when exposed to waterlogging stress (Duhan et al., 2017a, b). Similarly, an increase in the activity of SOD, CAT, POD, and APX was noticed when pigeon pea genotypes were exposed to waterlogging stress (Kumutha et al., 2009). Waterlogging also affects the metabolism of carbohydrates, which are the primary energy source in plants (Akhtar and Nazir, 2013). Waterlogging completely blocks oxidative phosphorylation, reducing available sugars such as glucose and fructose, which act as a source for ATP and allow glycolysis under conditions of hypoxia (Kuai et al., 2016; Duhan et al., 2017d). As such, plants have to maintain sufficient readily metabolizable sugars to survive in hypoxic or anoxic conditions (Sairam et al., 2009; Duhan et al., 2017d). Under hypoxic conditions, tolerant plants tend to maintain a higher sugar content (total, reducing, and non-reducing sugars) in their roots compared to sensitive ones (Hossain and Uddin, 2011). The development of gas-filled cavities (aerenchymas) in root cortical tissue also plays a pivotal role in the tolerance of plants to waterlogging. Aerenchyma formation is considered to be an adaptive response of plants against flooding (Drew et al., 2000; Evans, 2004) and is found in several crops like wheat, maize, pigeon pea, and sunflower. These gas-filled cavities act as a pathway for the movement of O_2 from the shoots to the roots (Hossain and Uddin, 2011).

Drought

A prolonged period of abnormally low rainfall leading to a shortage of water is the usual cause of drought. Drought is undoubtedly the biggest factor limiting plant growth and yields throughout the world. Tropical regions are now heavily affected by climate change and global warming, with increasing temperatures and concentrations of atmospheric CO_2, and decreasing water availability in the soil, affecting plant photosynthesis, development, and biomass (Kellomäki et al., 2005). Water plays an important role as an electron donor in photosynthesis meaning that its shortage can be a limiting factor (Angelopoulos et al., 1996; Nepomuceno et al., 1998). Low water availability also decreases stomatal conductance, which is the first strategy that plants use to decrease the transpiration rate and maintain turgor pressure (Eckstein et al., 1996). Drought induces decreased stomatal conductance, ultimately harming the reaction centres of photosystems I and II (PSI and PSII), by restricting proper carbon dioxide supply and leading to excessive loss of water vapour and decreased photosynthetic efficiency (Angelopoulos et al., 1996). In controlling their stomatal openings, plants reduce water loss by decreasing the transpiration flux and entry of CO_2 is simultaneously reduced. This plant response leads to ROS production under water deficient conditions (Mittler et al., 2002). Several studies have been conducted that show increased ROS accumulation in plants under drought stress (Beis and

Patakas, 2012). Water loss due to stomatal closure restricts CO_2 supply, or CO_2^- limited carbon fixation, and reduces $NADP^+$ regeneration throughout the Calvin cycle. Noctor (2000) reported increased photosynthetic electron transport and a reduced CO_2 fixation rate in the leaves of stressed plants. This imbalance between electron transport and the rate of CO_2 fixation results in a noticeable reduction in the electron transport chain and extra electrons are transferred onto O_2 molecules through the Mehler reaction (Asada, 1999). Biehler and Fock (1996) reported an approximately 50 % increase in the leakage of photosynthetic electrons through the Mehler reaction in drought stressed wheat plants in comparison to non-stressed plants. As CO_2 fixation is limited due to stomatal closure under drought stress, ribulose-1,5-bisphosphate is oxidized resulting in increased photorespiration (Noctor et al., 2002). This also results in ROS production amounting to 70 % of total H_2O_2 production due to photorespiration under drought stress conditions (Noctor et al., 2002). In addition to H_2O_2, other ROS like O_2^-, is also increased in stressed plants (Robinson and Bunce, 2002). This increased ROS production results in lipid peroxidation and protein modification in stressed plants (Sharma and Dubey, 2005; Esfandiari et al., 2008). The increased production of ROS leads to increased expression of genes for antioxidants and the levels of non-enzymatic and enzymatic antioxidants rise, resulting in greater tolerance to drought stress (Esfandiari et al., 2008). Sharma and Dubey (2005) studied the effect of water stress on rice plants and found increased *de novo* synthesis of MDHAR, DHAR, and GR, which is responsible for ASC and GSH regeneration. This is a primary response to water deficit. An increase in the activity of antioxidative enzymes has been reported in peanuts (Azevedo et al., 2010) and sugarbeet (Sayfzadeh and Rashidi, 2011) under water deficit conditions; this increase was found helpful in enhancing the capacity of the antioxidative system to scavenge ROS, thereby suppressing the level of lipid peroxidation in both plant species. Some plants show higher activity of antioxidative enzymes. This difference in tolerance can also be seen in different species of the same genus or even in cultivars of the same species. Plants showing greater antioxidant activity are described as tolerant and those that show comparatively lower activity are known as sensitive. Various studies have shown greater antioxidant capacities in drought tolerant plant genotypes compared to drought sensitive plants under drought conditions. Reddy et al. (2004) subjected five cultivars of the mulberry to drought stress and found two of them more tolerant; they also showed higher activity of antioxidative enzymes. Similarly, Moussa and Abdel-Aziz (2009) studied the effect of drought stress on two maize genotypes (one comparatively tolerant and one sensitive) and noticed increases in SOD, CAT, and POX activity in the tolerant genotypes; this was found to be helpful in maintaining lower levels of MDA and H_2O_2 in tolerant genotypes compared to sensitive ones. Sairam et al. (1998) found that a drought-tolerant wheat genotype exhibited greater APX and CAT activity and had higher ASC content and lower H_2O_2 and MDA in comparison to a drought-susceptible wheat genotype. When two genotypes of *Malus hupehensis* were subjected to drought stress, increases in H_2O_2, O_2^-, and MDA levels were observed in drought sensitive genotypes (*M. hupehensis*) compared to drought-tolerant ones (*M. prunifolia*). In contrast, SOD, POX, APX, GR, and DHAR activity and GSH contents increased, to a greater extent, in *M. prunifolia* than in *M. hupehensis* (Wang et al., 2012). Al-Ghamdi et al. (2009) reported that drought-acclimated leaves of wheat plants exhibit a systematic increase in APX and CAT activity and the maintenance of an adequate ascorbate redox pool through the efficient functioning of the APX enzyme, resulting in lower membrane damage in the drought-acclimated plants. The duration and severity of stress was also found to be a determining factor in the plant response. The drought response of a plant species depends on the duration and severity of the drought period. Marron et al. (2006) subjected two clones of *Populus deltoids* x *nigra* to drought stress and found increased SOD and CAT activity in mature leaves. For both clones, Mn-SOD, Fe-SOD, and Cu/Zn-SOD isoforms were detected in varying amounts, depending on drought intensity. On the basis of the above, it is clear that the antioxidative system plays an important role in plant tolerance of drought stress. Additionally, the phytohormone abscisic acid has also been found to play an important role in providing plant tolerance to drought stress. The roots send signals to the shoots through the xylem sap, which contains abscisic acid, further inducing stomatal closure in leaves and helping the plant to conserve water under conditions of water deficit (Davies et al., 1991).

Heavy metals

Metals, as cofactors for several enzymes, are involved in a number of biochemical reactions and play an important role as nutrients in the growth of plants. These elements are categorized into two categories on the basis of their use in plants, i.e. macronutrients and micronutrients. Metals and non-metals, like calcium (Ca), sulfur (S), and magnesium (Mg), are macronutrients and are used in various physiological activities like photosynthesis. Metals and other elements, like iron (Fe), boron (B), chlorine (Cl), manganese (Mn), zinc (Zn), copper (Cu), molybdenum

(Mo), and nickel (Ni), are micronutrients. In addition, metals serve as electron donors or acceptors in enzymes, which are needed for various parts of the metabolic cycle. For example, Cu, Zn, and Mn are used as cofactors of three isoforms (Cu, SOD, Zn SOD, and Mn SOD) of an important antioxidative enzyme SOD (Table 1). Zinc is used as a cofactor for carbonic anhydrase, another important enzyme that converts CO_2 to HCO_3^- and plays an important role in photosynthesis and stomatal closure in plant leaves. Copper is a multifunctional structural element that plays an important role in many processes such as photosynthetic electron transport, mitochondrial respiration, oxidative stress responses, cell wall metabolism, and hormone signaling (Raven et al., 1999). Cu is needed for the activity of several enzymes, including Cu/Zn SOD, cytochrome c oxidase, amino oxidase, laccase, plastocyanin, and polyphenol oxidase (Yruela, 2005) [**20**]. These metals are only needed in small amounts, but are necessary for life and growth in plants. Plants obtain these metals from soil, which contains various heavy metals, while others are added through various anthropogenic activities like mining, the smelting of metals, the burning of fossil fuels, the use of fertilizers and pesticides in agriculture, the industrial production of batteries and other metal products, sewage sludge, and municipal waste disposal. Not only do these activities make the soil barren and decrease plant yields, but they also see the entry of metals into plant and animal systems (Chibuike and Obiora, 2014). A study was conducted into the uptake and translocation of Cd, Cu, Fe, and Pb in different parts of three roadside plant species grown in the semi-urban zone of Jengka in the state of Pahang, Malaysia. It was found that metal content was greater in plants grown on contaminated sites compared to uncontaminated ones (Sulaiman et al., 2018). *Chromolaena odorata* plants were subjected to a nutrient solution containing three metals (Pb, Cd, and Zn) and metal accumulation was observed. Plants grown in an uncontaminated solution showed better growth compared to those grown in a contaminated solution (Tanhan et al., 2007). In addition to their use in important biochemical processes, in higher amounts these metals can become dangerous to plants. Heavy metals, including lead, nickel, cadmium, copper, cobalt, chromium, and mercury, are considered serious toxic environmental pollutants and decrease productivity. They have been identified as major stressors of plants. Heavy metals are taken up by plants in their soluble forms and used for growth. During the process of accumulation of essential metals some non-essential metals are also accumulated by plants (Djingova and Kuleff, 2000). Plants are unable to break down metals once their concentration is too high, increasing stress and having negative affects. Some of the consequences of heavy metal stress are: reduced photosynthetic activity; inhibition of cytoplasmic enzymes; damage to cell structures due to oxidative stress (Jadia and Fulekar, 2009); replacement of essential nutrients at cation exchange sites (Taiz and Zeiger, 2002); and a reduction in the number of beneficial soil microorganisms. This leads to a decline in soil nutrients due to a decrease in organic matter decomposition. Furthermore, soil microorganisms are involved in many useful enzymatic plant activities, which are also affected when soil microorganisms are reduced due to heavy metal stress. All of these toxic effects lead to a decline in plant growth and ultimately result in the death of the plant (Schaller and Diez, 1991). Different heavy metals have different effects on plants. Some metals, like Pb, Cd, Hg, and As, which are not needed for growth, are toxic at very low concentrations. A study recorded a significant reduction in height and reduced tiller and panicle formation in rice plants growing in soil contaminated with 1 mgHg/kg (Kibra, 2008). Ahmad et al. (2012) grew wheat plants in a soil solution with a Cd concentration of 5 mg/L and observed a reduction in shoot and root growth. Borkert et al. (1998) conducted experiments on peanut (*Arachis hypogaea* L.), soybean (*Glycine max* L.), corn (*Zea mays* L.), and rice (*Oryza sativa* L.) grown in mixtures of soils with varying levels of Zn and Cu. They observed that peanut and soybean were more sensitive to Zn toxicity, whereas corn and rice were more sensitive to Cu toxicity. The critical toxic level for peanut was 230 mg kg^{-1}, while for soybean it was 140 mg kg^{-1}, and for corn it was 20 mg kg^{-1}. Zhao and Duo (2015) also conducted an experiment to study the efficiency of native monocotyledonous weed species for heavy metal phytoextraction from municipal solid waste (MSW) compost and found that cadium (Cd), copper (Cu), and zinc (Zn) accumulated inside the plants and affected growth. Plants have also developed defense mechanisms in order to survive these stress conditions. This process involves the synthesis and activation of stress-related proteins and the activation of various signaling molecules. Many transcription factors are activated that induce the transcription of various metal stress-response genes as a result of signal transduction. The various signaling pathways activated in response to heavy metal stress are: the MAPK cascade; hormonal signaling; the ROS signaling pathway; and the Ca–Calmodulin pathway. Restricting the uptake of metal ions via roots is one mechanism plants use as a first line of defense. Other mechanisms include the targeting of specifc metal ions, sequestering and excluding them into compartments and preventing the exposure of other cellular organelles. Some adaptive mechanisms are also activated only once a stressor is encountered. Most metals enter plant roots through energy-driven pumps or ion channels. Plants restrict the uptake of metal ions by changing the pH of the rhizosphere resulting in the precipitation of heavy metals. Plant root exudates cause pH changes allowing these metals ions to be concentrated in the apoplast rather than entering the cell. Pinto et al. (2008) found that root exudates containing malate were found helpful in lessening the susceptibility of sorghum bicolor to cadmium toxicity. Sanita Di Toppi and Gabbrielli (1999) also found oxalate secretion helpful in restricting cadmium uptake in tomato plants. The production of a specific root exudate was found to particularly characterise tolerant plants. The secretion of a high quantity of root exudates containing malic acid was observed in all tolerant genotypes of wheat exposed to aluminum toxicity (Huang et al., 1996). Root exudates nullify the effect of metal ions by binding them to the cell wall in order to restrict their movement and accumulate them between the cell wall and the plasma membrane of the root cells. In addition to this, an iron plaque has been seen to form on the root surface, particularly in rice plants, releasing oxygen and oxidants into the rhizosphere (Chen et al., 1980). It has been proposed by Liu et al. (2008) that this iron plaque is responsible for sequestering and absorbing cadmium in the soil, increasing plant resistance. Citric acid has a strong binding afinity for heavy metals, especially towards iron, and a comparatively lower afinity for Cd, Ni, Co, and Zn. Similar sequestering and chelating effects of malate and oxalate exudates have also been observed in other plants. In addition, histidine and NO have been reported to chelate heavy metals in plant cells and the xylem (Kramer et al., 1996). NO is also involved in plant responses to heavy metal stress as a signaling molecule that regulates the level of hormones and ROS. NO can be both beneficial and harmful depending upon its concentration and the location of the NO in plant cells (Ozturk et al., 2013; Hakeem et al., 2017). It is responsible for the conversion of oxygen radicals to hydrogen peroxide and oxygen, reducing oxidative stress and enhancing the activity of enzymes that bind and scavenge hydrogen peroxide. NO itself also scavenges free radicals (Lamattina et al., 2003; Shi et al., 2007; Zheng et al., 2009; Siddiqui et al., 201). Xiumin et al. (2009) applied NO externally to tomato plants growing under copper stress. NO alleviated the stress of the tomato plants and was seen to induce the activity of ATPase and PPase in the plasma membrane, as well as in the tonoplast of the vacuole. These results also suggest that NO is responsible for providing heavy metal tolerance to plants. NO is also responsible for increasing calcium and the calcium–calmodulin pathway is involved in providing heavy metal tolerance. Calcium is a secondary messenger that stimulates the protein calmodulin, which, in turn, regulates heavy metal uptake, transport, and metabolism. The Ca–calmodulin pathway has been studied in relation to Cd, Ni, and Pb stress, and it is strongly suggested that calmodulin is an important cellular target for various metals. NO is reported to be involved in increasing the cytosolic concentration of Ca^{2+} in response to heavy metal and other abiotic stresses. Elevated levels of calcium cause physiological and biochemical changes in the cell along with elevation or regulation of NO concentration and thus they have a synergistic role in abiotic stress control (Khan et al., 2014). Heavy metal stress also leads to ROS production and causes oxidative stress in plants (Jalmi et al., 2018). Reactive oxygen species (H_2O_2, superoxide radicals, and hydroxyl radicals) are formed due to a reduction in oxygen caused by heavy metal stress. ROS cause the peroxidation of lipids, proteins, DNA, and cellular membranes. In addition to being harmful, ROS control various physiological processes, e.g. germination, stomatal closure, and senescence (Yuan et al., 2013). Reactive oxygen species become harmful when produced in high amounts and antioxidants, such as ascorbic acid, tocopherol, and glutathione, and enzymes, including SOD, CAT, GR, POX, and APX, are produced to regulate their concentration (Sytar et al., 2013). Of all the reactive oxygen species, H_2O_2 is the only one that can cross the plasma membrane; it is directly involved in cell-to-cell signaling (Mittler et al., 2004; Pitzschke et al., 2006). Mitogen-activated protein kinases (MAPKs) are also involved in heavy metal stress signaling. They are involved in the production of phosphatidic acid, which is also an important secondary messenger molecule and has a role to play in the synthesis of ROS (Bargmann and Munnik, 2006). Plant hormones and phenolics are involved in heavy metal stress. Phytohormones like Jasmonic acid and ethylene is known as stress hormones and plays a pivotal role during stress. In addition, the concentration of salicylic acid is also reported to increase during metal stress. Hossain et al. (2012) found salicylic acid to have a role in regulating ROS and antioxidant concentrations. An experiment showing the formation of an auxin-responsive mRNA in response to Cd stress was conducted with *Brassica juncea* (Minglin et al., 2005). Shao et al. (2010) observed mercury induced synthesis of ABA and salicylic acid during proteomic analysis of *Arabidopsis* and pea plants. Cadmium was also shown to induce the production of ethylene and abscisic acid (Polle and Schutzendubel, 2004). The accumulation of the amino acid proline has also been observed in plants growing under heavy metal stress, especially in heavy metal-tolerant plants. In fact, tolerant plants accumulate high levels of proline even in the absence of metal stress. Proline has been reported to be involved in neutralizing the free radicals generated from exposure to heavy metals, but their role in sequestering the metals has not been identified. Hossain et al. (2010) reported exogenous proline induced tolerance against cadmium stress in mung beans. Polyamines, like spermidine, spermine, and putrescine, have also been found to improve tolerance to heavy metal stress (Hossain et al., 2012). Phenolics, which are secondary metabolites, have also been found to be synthesized during heavy metal stress. These compounds have at least one aromatic ring and one or more hydroxyl groups attached. They are synthesized from cinnamic acid and a number of studies have shown their synthesis to be enhanced in plants exposed to heavy metal stress. When sprayed with copper sulfate, *Phyllantus tenellus* showed increased accumulation of phenolics in its leaves (Diáz et al., 2001). Similarly, when exposed to cadmium, *Phaseolus vulgaris* accumulated soluble and insoluble phenolics (Winkel-Shirley, 2002). The same has been observed in wheat in which the synthesis of phenolics greatly increases in response to Ni stress (Diáz et al., 2001). The phenolics possess antioxidant properties that chelate metals. The hydroxyl group present in them binds to metals, particularly Cu and Fe. Phenolic compounds inhibit lipid peroxidation by trapping lipid alkoxyl radicals produced during stress conditions. The flavonols—a group of phenolics—have protein-binding capabilities and can anchor phospholipids through hydrogen bonding. This enables them to accumulate near the cytosolic and extracellular side of the lipid membrane. In this way, they also maintain the integrity of bound membranes. Phenolics, being antioxidants, readily scavenge free radical species, such as H_2O_2 and superoxide ions.

20. Metals acting as cofactors for enzymes involved in various important metabolic reactions.

Name of metal ions	**Enzyme containing this ion**	**Function of enzymes**
Cupric ion	Cytochrome oxidase	Cytochrome c oxidase is a large trans-membrane protein complex found in bacteria, archaea, and the mitochondria of eukaryotes. It receives an electron from each of four cytochrome c molecules, and transfers them to one dioxygen molecule, converting the molecular oxygen into two molecules of water.
	Superoxide dismutase	SOD is an enzyme that alternately catalyzes the dismutation of the superoxide (O_2^-) radical into ordinary molecular oxygen (O_2) and hydrogen peroxide (H_2O_2). Cu-SOD is localized in the cytosol, chloroplasts, peroxisomes, and apoplast.
Ferrous or Ferric ion	Catalase	Catalase is a common enzyme found in nearly all living organisms exposed to oxygen (such as bacteria, plants, and animals). It catalyzes the decomposition of hydrogen peroxide into water and oxygen.
	Cytochrome	Cytochromes are redox-active proteins containing a heme with a central Fe atom at its core as a cofactor. They are involved in the electron transport chain and redox catalysis.
	Nitrogenase	Nitrogenases are enzymes that are produced by certain bacteria, such as cyanobacteria (blue-green bacteria). These enzymes are responsible for the reduction of nitrogen (N_2) to ammonia (NH_3).
	Hydrogenase	A hydrogenase is an enzyme that catalyzes the reversible oxidation of molecular hydrogen (H_2).
	Superoxide dismutase	SOD is an enzyme that alternately catalyzes the dismutation of the superoxide (O_2^-) radical into ordinary molecular oxygen (O_2) and hydrogen peroxide (H_2O_2). Fe-SOD has been found mainly in chloroplasts, but has also been detected in peroxisomes
Manganese	Arginase	Arginase is a manganese-containing enzyme. The reaction catalyzed by this enzyme is: arginine + $H_2O \rightarrow$ ornithine + urea. It is the final enzyme of the urea cycle and is ubiquitous to all domains of life.

	Superoxide dismutase	Mn-SOD is present in mitochondria and peroxisomes. SOD is an enzyme that alternately catalyzes the dismutation of the superoxide (O_2^-) radical into ordinary molecular oxygen (O_2) and hydrogen peroxide (H_2O_2).
Molybdenum	Nitrate reductase	Nitrate reductases are molybdoenzymes that reduce nitrate (NO^-_3) to nitrite (NO^-_2).
	Nitrogenase	Nitrogenases are enzymes that are produced by certain bacteria, such as cyanobacteria (blue-green bacteria). These enzymes are responsible for the reduction of nitrogen (N_2) to ammonia (NH_3).
Nickel	Urease	Ureases are found in numerous bacteria, fungi, algae, plants, and some invertebrates. These enzymes catalyze the hydrolysis of urea into carbon dioxide and ammonia.
Zinc	Alcohol dehydrogenase	Alcohol dehydrogenases (ADH) are a group of dehydrogenase enzymes that occur in many organisms and facilitate inter-conversion between alcohols and aldehydes or ketones with the reduction of nicotinamide adenine dinucleotide (NAD^+) to NADH.
	Carbonic anhydrase	The carbonic anhydrases or carbonate dehydratases form a family of enzymes that catalyze inter-conversion between carbon dioxide and water and the dissociated ions of carbonic acid (i.e. bicarbonate and hydrogen ions).
	DNA polymerase	A DNA polymerase is a member of a family of enzymes that catalyze the synthesis of DNA molecules from nucleoside triphosphates, the molecular precursors of DNA. These enzymes are essential for DNA replication.
	Superoxide dismutase	SOD is an enzyme that alternately catalyzes the dismutation (or partitioning) of the superoxide (O_2^-) radical into ordinary molecular oxygen (O_2) and hydrogen peroxide (H_2O_2). Zn-SOD has been localized in the cytosol, chloroplasts, peroxisomes, and apoplast.

High temperature stress

Temperature is also a determining factor for plant growth. The global air temperature is rising by 0.2 °C per decade, which will lead to temperatures 1.8-4.0 °C higher than the current level by 2100. From this prediction, we can conclude that high temperatures may well become life threatening in the near future as the effect on plant growth and yield is already well known. The inability of plants to move from unfavourable conditions makes them more susceptible to stress. Heat stress affects plant growth, development, and yield. At the growth stage, germination is the first process to be affected by heat stress. Major consequences of heat stress include: reduced germination, lower plant emergence, abnormal seedlings, poor seedling vigor, and reduced radicle and plumule growth in geminated seedlings (Kumar et al., 2011; Piramila et al., 2012). ABA is considered a major reason for the Inhibition of seed germination (Essemine et al., 2010). High temperatures also impact photosynthesis (Crafts-Brandner and Salvucci, 2002). C_3 plants are more prone to damage than C_4 plants (Crafts-Brandner and Salvucci, 2002) and primary sites for injury are located in the chloroplast. Processes that are primarily affected include the carbon metabolism of the stroma and photochemical reactions in thylakoid lamellae (Yang et al., 2006; Wang et al., 2009). The thylakoid membrane is highly susceptible to high temperature and significant alterations take place under heat stress, such as changes to the structural organization of thylakoids, loss of grana stacking, and swelling of grana (Rodríguez et al., 2005). Photosystem II is also affected by high temperatures (Morales et al., 2003). In addition, changes like decreased leaf water status, leaf stomatal conductance, and intercellular CO_2 concentration all lead to decreased photosynthesis through heat stress (Greer and weedon, 2012). Heat stress induces closure of stomata also resulting in decreased photosynthesis (Ashraf and Hafeez, 2004). Plants with tolerance to heat stress try to maintain normal rates of gas exchange and CO_2 assimilation even under hot conditions (Kumar et al 2005; Yang et al 2006). Photosynthetic pigments are also lowered (Marchand et al., 2005). High temperatures have been shown to induce lipid peroxidation of chloroplasts and thylakoid membranes resulting in decreased chlorophyll pigment in sorghum (Djanaguiraman et al., 2010). Djanaguiraman et al. (2010) found that heat stress (38/28 °C) significantly decreased total chlorophyll content (18 %), chlorophyll *a* content (7 %), chlorophyll *a*/*b* ratio (3 %), and leaf stomatal conductance (16 %) in soybean plants. This further resulted in decreased sucrose content (9 %) and reduced sugar content (4 %) and leaf soluble sugar content (36 %) (Djanaguiraman et al., 2011). In rice plants, high temperatures (33 °C, 5 days) decreased the photosynthetic rate by 16 % in the Shuanggui 1 variety and 15 % in T219 (Cao et al., 2009). Greer and Weedon (2012) observed that average rates of photosynthesis of *Vitis vinifera* leaves decreased by 60 % with an increase temperature from 25 to 45 °C. This reduction in photosynthesis was attributed to 15–30 % stomatal closure. Enzymes, like sucrose phosphate synthase, ADP-glucose pyrophosphorylase, and invertase, are also affected by heat stress (Sumesh et al., 2008). Kumar et al. (2005) considered reduced leaf water potential, reduced leaf area, and pre-mature leaf senescence under heat stress responsible for decreased photosynthesis. Plants are more susceptible to heat stress at the reproductive stage compared to the vegetative one (Mittler and Blumwald, 2010). Even a short period of heat stress can cause a significant decrease in floral buds and aborted flowers at the reproductive stage (Young et al., 2004). During its reproductive developmental stages, heat may cause plants to produce no flowers or flowers that cannot produce fruit or seeds (Sato et al., 2006). Heat stress induces impaired meiosis in both male and female organs; impaired pollen germination and pollen tube growth; reduced ovule viability; anomalies in stigmatic and style positions; a reduced number of pollen grains retained by the stigma; disturbed fertilization processes; and problems in the growth of the endosperm, proembryo, and unfertilized embryo, which are all major causes of sterlity (Foolad et al., 2005). Cao et al. (2008) subjected rice plants to heat stress (>33 °C) at the heading stage and observed a reduction in anther dehiscence and the pollen fertility rate, leading to a reduction in the number of pollen grains on the stigma. This causes reduced fertilization, lower spikelet fertility, and sterile seeds; more damage is seen in sensitive varieties compared to tolerant varieties. Heat stress (32 °C) was shown to result in increased spikelet sterility (by 61 % compared to control) due to decreased pollen germination (36 %) in rice when applied during the night (Mohammed and Tarpley, 2010). High temperature has been shown to induce excessive production of ethylene, which is considered a cause of male sterility in rice pollens. Ethylene is hypothesized to inhibit the key enzymes in sugar–starch metabolism, which weakens sink strength and restricts grain filling, ultimately producing sterile grain. High temperature stress was seen to result in the abscission and abortion of flowers, young pods, and developing seeds, resulting in lower seed numbers in soybean (Tubiello et al., 2007). High temperatures during flowering are known to decrease pollen viability in soybean (Djanaguiraman et al., 2011). Edreira and Otegui (2012) observed that heat stress during flowering periods, specifically at the pre-silking and silking stages, resulted in higher yield reduction relative to heat stress at the grain filling stage of maize. Various crops including cereals showed heat stress induced a reduction in yield of grains (e.g., rice, wheat, barley, sorghum, maize), pulses (e.g., chickpea, cowpea), oil crops (mustard, canola), and so on (Warland et al., 2006; Cao et al., 2008; Edreira and Otegni, 2012). It was demonstrated that an increase in the seasonal average temperature by 1 °C decreased the grain yield of cereals by 4.1 to 10.0 % (Hatfield et al., 2011). Cao et al. (2008) observed the effect of heat stress on sensitive and tolerant varieties of rice and observed that 1,000-grain weight was reduced by 7.0-7.9 % in the sensitive Shuanggui 1 and 3.4-4.4 % in the tolerant Huanghuazhan varieties of rice. A significant yield reduction was also observed in the heat-sensitive rice cultivar Shuanggui 1 (35.3 to 39.5 %) compared to the heat tolerant cultivar Huanghuazhan (21.7 to 24.5 %). Mohammad et al. (2010) observed that high night temperatures (32 °C) resulted in decreased grain length (2 %), width (2 %), and weight in *O. sativa*, along with increased spikelet sterility (61 %). It also increased grain nitrogen (N) concentration (44 %), which was inversely related to grain weight. All of these factors contribute to a reduction in yield (90 %). The main cause for yield loss is reduced photosynthesis due to membrane disintegration, enhanced maintenance respiration costs, and a reduction in radiation use efficiency under heat stress. Gunawardhana et al. (2011) observed damage in pod quality parameters, such as fibre content, and the break down of calcium pectate was found in okra (*Abelmoschus esculentus*). Heat stress is also thought to uncouple enzymes and metabolic pathways, causing the accumulation of unwanted and harmful ROS, most commonly singlet oxygen (O_2), superoxide radicals ($O_2^{\cdot -}$), H_2O_2, and hydroxyl radicals ($^{\cdot}OH$), which are responsible for oxidative stress (Vasseur et al., 2011). Though ROS are also generated in other organelles, i.e. peroxisomes and mitochondria, the reaction centers of photosystem I and photosystem II in chloroplasts are considered to be the major sites of ROS generation (Asada, 2006). Hydroxyl radicals can potentially react with all biomolecules, including pigments, proteins, lipids, and DNA, and with almost all cell constituents (Halliwell, 2006). Singlet oxygen can directly oxidize proteins, polyunsaturated fatty acids, and DNA (Karuppanapandian et al., 2011). Thermal stress can induce oxidative stress through peroxidation of membrane lipids and disruption of cell membrane stability by denaturation of proteins (Huang and Xu, 2008). Cao et al. (2009) observed increased malondialdehyde (MDA) content in rice plant leaves as a result of heat stress. Bavita et al. (2012) observed how heat stress (33 °C) induced damage to membrane properties, protein degradation, and enzyme deactivation in wheat, which significantly reduced cell viability. Heat stress induces oxidative stress, significantly increases membrane peroxidation, and reduces membrane thermostability by 28 % and 54 %, which, surprisingly, increases electrolyte leakage in wheat. The modification of physiological and biochemical processes by changes to gene expression gradually leads to the development of heat tolerance through acclimation, or, in the ideal case, adaptation (Semenov and Halford, 2009; Moreno and Orellana, 2011). Under heat stress, plants use many avoidance mechanisms, such as changing leaf orientation; transpirational cooling; alteration of membrane lipid compositions; closure of stomata, reducing water loss; increased stomatal and trichomatous densities; and larger xylem vessels. Early maturation in many crop plants is closely correlated with smaller yield losses under heat stress as an effective escape mechanism. Plants growing in hot climates avoid heat stress by reducing the absorption of solar radiation. They have developed small hairs that form a thick coat on the surface of the leaf, as well as cuticles—a protective waxy covering. Plants use strategies, such as turning leaf blades away from direct sunlight, orienting themselves parallel to the sun's rays (paraheliotropism), and rolling of leaf blades to avoid excess absorption of light. In addition, many enzymatic and nonenzymatic antioxidants are synthesized. Studies in *Lens culinaris* have observed an increase in the activity of CAT, APX, and SOD before declining at 50 °C, while POX and GR activity was seen to decline at all temperatures ranging from 20 to 50 °C. In addition, total antioxidant activity was seen to reach its maximum at temperatures of 5 to 40 °C in tolerant varieties and at 30 °C in susceptible ones (Chakrabortty and Pradhan, 2011). Sairam et al. (2000) found antioxidant metabolites, like AsA, GSH, tocopherol, and carotene, helped mitigate oxidative stress. Xu et al. (2006) found lower production of ROS in turf grass as a result of enhanced synthesis of AsA and GSH. Kocsy et al. (2000) observed how heat stress raised GSH levels and increased the activity of enzymes involved in GSH synthesis and the GSH/GSSG ratio in wheat. Heat shock proteins (HSPs) are also found to mitigate heat stress. They are grouped into five different families: HSP100, HSP90, HSP70), HSP60, and HSP 20 (or small HSP, sHSP) (Swindell, 2007). The HSP70 and HSP60 proteins are among the most highly conserved proteins in nature, consistent with a fundamental role in the heat stress response (Kulz, 2003). Besides this, the exogenous applications of protectants in the form of osmoprotectants (proline, glycine betaine, trehalose etc.), phytohormones (abscisic acid, gibberellic acid, jasmonic acid, brassinosterioids, salicylic acid, etc.), signaling molecules (e.g. NO), polyamines (putrescine, spermidine, and spermine), trace elements (Se, Si, etc.), and nutrients (N, P, K, Ca, etc.) have been found effective in mitigating high temperature stress induced damage in plants (Hasanuzzaman et al., 2011; Hasanuzzaman et al., 2012; Hasanuzzaman et al., 2013; Waraich et al., 2012).

Low temperature stress

Temperature stress, especially cold stress, has been reported to have very wide and far-reaching consequences for many agricultural and horticultural crops. Due to the sessile and poikilothermic nature of plants, they can expect to face greater temperature extremes in the near future brought about by global climate change. Low temperature stress has two types: chilling stress and freezing stress. Chilling stress occurs when temperatures go below 10 to 15 °C, whereas freezing stress occurs when temperatures go below 0 °C and ice forms within tissues. Chilling damage can cause many irreversible changes in metabolic processes, leading to long-duration aberrations. Cellular changes, like changes to membrane structure and composition, decreased protoplasmic streaming, electrolyte leakage, and plasmolysis, take place in response to chilling stress. Other changes, like increased or reduced respiration depending on the severity of stress, or the production of abnormal metabolites due to anaerobic conditions, also take place in response to chilling stress. Symptoms of chilling stress include: reduced plant growth and death; surface lesions on leaves and fruits; abnormal curling; lobbing and crinkling of leaves; water soaking of tissues; cracking, splitting, and dieback of stems; internal discolouration (vascular browning); increased susceptibility to decay; failure to ripen normally; loss of vigour; and finally wilting followed by water soaked patches that develop into sunken pits and later dry up, leaving necrotic patches of tissues on the leaf surface. Fruits show somewhat different symptoms, for example sunken pits in cucumbers, browning of skin and degradation of pulp tissue in bananas, and blackened hearts in pineapples (Wilson, 1987). Chilling stress can induce a phase transition in the lipid bilayer, from a liquid crytalline phase to a solid gel state, which results in cell membrane damage, further leading to leakage of organic and inorganic substances. Lipid composition plays a pivotal role in tolerance to chilling stress—the ratio of unsaturated to saturated fatty acids is higher in chill-resistant plants due to the increase in activity of fatty acid desaturases.

Freezing injury in plants can be from two sources: 1. freezing of soil water; 2. freezing of the fluids within the plant. Soil water available to plants is found in the porous regions between soil particles. It freezes at about -2 °C, depriving the plant of its main source of water. Freezing of water within the plant is a more serious threat, as it can cause disruption to the structure and function of cells and tissues. Freezing damage occurs primarily due to the formation of ice crystals, which damage cell structure when the temperature falls below 0 °C. Ice usually forms first in the cell walls and intercellular spaces; damage occurs when ice crystals grow and puncture the cytoplasm (Xin and Browse, 2000). In addition to the above changes, oxidative stress caused by ROS is also a consequence of cold stress (Guo et al., 2006). However, to counter chilling and freezing stresses, plants have developed diverse defense mechanisms, including the accumulation of carbohydrates and proteins, resulting in large quantities of soluble sugars, amino acids, and cold induced stress-related proteins (Trischuk, 2014); furthermore, hormone homeostasis induces gene expression, which functions to stabilize membranes against freezing induced injury (Fowler and Thomashow, 2012). Plants exposed to low temperatures show changes in concentrations of water-soluble carbohydrates, like monosaccharides (glucose and fructose), disaccharides (sucrose and trehalose), and fructans (Theocharis et al., 2012a), which act as osmoprotectants, cellular membrane stabilizers, scavengers of reactive oxygen species, and signaling molecules, making them more tolerant of chilling and freezing (Theocharis et al., 2012b). A number of studies have reported the role of higher accumulation of sugars, or individual sugar fractions such as starch (Zhang et al 2006), fructans (Dionne et al., 2001), and sucrose (Zhang et al., 2006), in cold acclimation. Studies conducted with turf grass also found higher accumulation of sugars helped make plants tolerant to cold stress (Cai et al., 2004; Hoffman et al., 2010). Some plant hormones are involved in regulating plant growth, development, and adaptation (Srivastava, 2002). Abscisic acid, in addition to its role in leaf senescence, seed dormancy and germination, and cell division and elongation, has also been found to be involved in regulating chilling and freezing in plants (Finkelstein and Gibson, 2002). Other hormones, like cytokinins, have also been found to have a role in delaying both stomatal closure and leaf senescence under abiotic stresses (Ryu and Cho, 2015). Jasmonic acid (JA) has also been identified as signaling molecule for defense responses in plants (Lianes et al., 2016) and has been shown to induce a response similar to that of abscisic acid to alleviate chilling injury (Hu et al., 2013). Salicylic acid (SA) has also been demonstrated to be involved in plant responses to a broad range of abiotic stresses, including low temperature (Khan et al., 2015). Some stress-related genes, such as the dehydration-responsive elements or C-repeat binding factor genes (*CBF/DREB1*), as well as the cold induced stress proteins, such as late-embryogenesis abundant (LEA) proteins (Hundertmark and Hincha 2008), are involved in cold acclimation. Low temperatures have been shown to induce upregulation of expression of *CBF/DREB1*, which, in turn, induces the expression of *COR* genes to enhance freezing tolerance (Kurepin et al., 2011). The antioxidative system was also found to be effective in developing tolerance of plants against low temperature stress. Guo et al. (2006) found higher AsA content associated with higher antioxidative capacity and higher cold tolerance in rice. Fortunato et al. (2010) experimented with coffea sp. and found that ROS levels (mainly H_2O_2 and OH^-) were reduced by the over production of AsA and α-tocopherol contents under low temperature stress. The ratio of GSH/GSSG is considered important because a higher ratio is an indication of better tolerance to stress—a higher GSH/GSSG ratio is needed to maintain a sufficient amount of GSH in the AsA-GSH cycle (Kocsy et al., 2000). Takac et al. (2003) found the activity of some antioxidant enzymes partially correlated with the chilling tolerance of maize cultivars. Huang and Guo (2005) found the antioxidative system more efficient in chill-tolerant cultivars than chill-susceptible ones. Wang and Li (2006) found the exogenous application of a salicylic acid (SA) pretreatment helped maintain higher APX, GR, and MDHAR activity and redox ratio in the AsA–GSH pool under both normal and cold stress conditions. They suggest that $Ca2^+$ homeostasis and antioxidant systems are behind SA induced cold tolerance. Zhao et al. (2009) also observed an increase in the activities of CAT, APX, POX, and SOD in tomato plants when subjected to low temperature stress. Reduced activity of antioxidant enzymes, i.e. SOD, POD, CAT, and APX, were observed in *Cucumis sativus* plants under low temperature stress. However, exogenous application of putrescine and spermidine helped to restore antioxidative enzyme activity, which rendered the plants more tolerant to chilling (Zhang et al., 2009). Liu et al. (2011) found that pretreatment of 1.0 mM SNP (NO donor) enhanced the activity of enzymes, such as SOD, GR, POD, and CAT, in *Cucumis sativus* seedlings under chilling conditions (4 °C), indicating that exogenous NO can enhance tolerance. It was also observed that MDA content was significantly decreased (27 %) in chilling stressed seedlings due to SNP treatment compared to stress alone. Yang et al. (2011) found enhanced activity of SOD, CAT, APX, and POX to be behind better tolerance of low temperature stress in *Cucumis sativus* plants. Turan and Ekmekci (2011) also found that cold-acclimated *Cicer arietinum* plants showed higher activity of SOD, APX, GR, and POX under a chilling stress of 2 and 4 °C, which enhanced their tolerance.

Salinity

Salinity is a major stressor affecting plant growth and productivity (Munns 1993; Gama et al. 2007). About 20 % of the world's cultivated land and approximately half of all irrigated land is currently affected by salinity (Zhu, 2001). According to the FAO Land and Plant Nutrition Management Service (2008), more than 800 Mha of land throughout the world is affected by salinity, amounting to more than 6 % of the world's total land area. Human induced salinity, also known as secondary salinization, due to improper cultivation practices and irrigation, is now a major problem for agriculture (Pannell and Ewing, 2004). For example, the use of brackish water for irrigation is a significant cause of soil salinization. Across the world, more than 45 million hectares of irrigated land have been damaged by excess salt and a further 1.5 million hectares are lost each year as a result of high levels of soil salinity (Munns and Tester, 2008). Salinity stress affects plant growth and productivity with adverse effects on germination, plant vigour and crop yield (Munns and Tester, 2008). Water stress, ion toxicity, nutritional disorders, oxidative stress, alteration of metabolic processes, membrane disorganization, and the reduction of cell division and expansion, are all consequences of high salinity (Munns, 2002; Zhu, 2007). Parida and Das (2005) undertook experiments on the initial phase of salinity stress and found that major processes such as photosynthesis, protein synthesis, and energy and lipid metabolism are all affected. Water stress is also experienced by plants during their first exposure to salinity stress in turn causing leaf wilting. Stomatal closure is also seen during the early phases of salinity stress (Munns, 2002). If salinity stress persists for a long duration, plants experience ionic stress resulting in premature senescence of adult leaves and leading to a reduction in photosynthetic area (Cramer and Nowak, 1992). Due to the capacity of sodium and, more importantly, chloride to affect plant enzymes, they can experience cell swelling, resulting in reduced energy production and other physiological changes during salinity stress. Ionic stress results in premature senescence of older leaves and in symptoms of toxicity (chlorosis, necrosis). In mature leaves, due to high Na^+ concentration, the disruption of protein synthesis and interference in enzyme activity have all been observed (Munns, 2002). Khan and Weber (2008) noticed that higher levels of salt stress inhibit seed germination. A decrease in tissue chlorophyll content with increased salinity was reported in oat plants (Zhao et al., 2007). This reduction in chlorophyll content is due to ion accumulation through inhibition of synthesis or increased disintegration of chlorophyll in leaves. The underlying cause of chlorophyll reduction is ion accumulation in leaves (Meloni et al., 2003). Sreeni-Vasulu et al. (2000) observed membrane damage in foxtail millet (*Setaria italica*) seedlings due to salinity stress. Membrane damage further results in electrolyte leakage, which can cause peroxidation of lipids, proteins, and nucleic acids in the cell. Quan et al. (2004) found a positive correlation between electrolyte leakage in sugar beet plants and salinity stress. Sevengor et al. (2011) experimented with pumpkin plants exposed to salinity stress and observed an increase in MDA content in all the studied genotypes, with low MDA and better growth in tolerant genotypes, Iskenderun-4 and AB-44 (Iskenderun-4, 53.03 % and AB-44, 34.65 %), compared to sensitive genotypes, CU-7 and A-24 (132.82 % and 140.48 %). Salt stress significantly reduced the number of flowers, number of pods, 100 seed weight, and number of seeds per pod. Salinity stress was also found to reduce the number of pods and grains per plant, grain weight, and grain yield in soybean (Ghassemi-Golezani et al., 2009). In addition, generation of ROS, such as superoxide anions (O_2^-), H_2O_2, and hydroxyl radicals, particularly in chloroplasts and mitochondria, is an important consequence of salinity stress in plants (Parida and Das, 2005; Zhang et al., 2013). Accumulation of H_2O_2 will also see enhanced potential for the production of hydroxyl radicals, leading to lipid peroxidation and membrane deterioration. These ROS can affect the integrity of cellular membranes, enzyme activity, and the photosynthetic apparatus of the plant (Jithesh et al., 2006). Osmotic adjustment acts as the first line of defense in response to salinity stress. For example, Duhan et al. (2018) reported accumulation of proline in pigeon pea plants exposed to 30 Mm salinity. Proline accumulation has also been observed in crops like chickpea (Kukreja et al., 2005), pearl millet (Mukhopadhyay et al., 2007), faba bean (Ahmed et al., 2008), Catharanthus (Jaleel et al., 2008), barley (Khosravinejab et al., 2009), mung bean (Saha et al., 2010), citrus (Abadi et al., 2010), pepper (Chookhampaeng, 2011), and durum wheat (Kahirizi et al., 2012). Duhan et al. (2017c) reported an increase in total soluble sugars in pigeon pea roots. The increase was greater in tolerant genotypes (ICPH-2431, PARAS) compared to sensitive ones (UPAS-120, H09-33). Amini and Ehsanpour (2005) reported similar results in two tomato cultivars, with sugars content increasing from 55.5 to 329 mg g^{-1} dry weight for 0 and 160 mM of NaCl, respectively. A decrease in soluble sugar content was observed in *Saueda salsa* with an increase in the salinity level (Duan et al., 2007). Majid et al. (2012) reported that soluble sugar concentration varied with the intensity of salt in three alfalfa varieties. On the basis of these studies, total soluble sugar can also be considered a determining factor for plant tolerance to salinity stress. In addition, antioxidant enzymes scavenge or prevent the formation of ROS acting as line of defense against salinity stress (Unal et al., 2014). There have been reports of increased activity of antioxidative enzymes under salt stress in *Hordeum vulgare* (Unal et al., 2014), *Zea mays* (Rios and Pinto, 2014), *Nicotiana tabacum* (Celik and Atak, 2012), and *Glycine max* (Aksoy and Dinler, 2012). Duhan et al. (2017a) exposed pigeon pea plants to 30mM salinity and observed increased activity of antioxidative enzymes, such as SOD, catalase, peroxidase, ascorbate peroxidase, glutathione reductase, dehydroascorbate reductase, and malon dehydroascorbate reductase. Additionally, nonenzymatic antioxidative metabolites, like ascorbate and glutathione, were found to be raised in all the genotypes. A greater increase was seen in tolerant genotypes (ICPH-2431, PARAS) compared to sensitive genotypes (UPAS-120, H09-33), indicating the role of the antioxidative metabolism in developing plant tolerance of salinity stress. Increased activity of antioxidative enzymes under salinity stress were also observed by Amirjani (2010) in *Glycine max*; Xiaoli et al. (2009) and Mahmood (2011) in *Hordeum vulgare*; Rios and Pinto, (2014) in *Zea mays*; and Celik and Atak (2012) in *Nicotiana tabacum*. Carbohydrate metabolism appears to play a role in increasing plant tolerance to salinity stress. Duhan et al. (2017c) studied four pigeon pea genotypes under 30 mM salinity stress and found increased activity of carbohydrate metabolism enzymes, i.e. sucrose synthase and alcohol dehydrogenase, in all four genotypes (ICPH-2431, PARAS, UPAS-120, and H09-33). The increase was higher in tolerant genotypes (ICPH-2431, PARAS) compared to sensitive genotypes (UPAS-120, H09-33). This may indicate the role of carbohydrate metabolism in developing tolerance of salinity stress.

Conclusions and future perspectives

Today, abiotic stress is becoming major threat to plant growth and yield. All abiotic stress, whether it is caused by waterlogging, salinity, drought, or heavy metals, is harmful to plants. However, the extent of damage depends upon a number of factors, including the duration of the stress, the genotype of the plant, and the stage at which the stress ocurrs. Plants can usually withstand stressful conditions for short durations, but if the duration exceeds its limits death will result. Depending on their response to stress, plants can be divided into two types—tolerant and sensitive. Tolerant plants show greater ability to cope with stress conditions compared to sensitive ones. Stress at later stages of growth has been found to cause more damage compared to early stages of growth. All types of stress cause cellular changes, such as chlorophyll degradation, excessive loss of water, reduced water uptake, cell membrane disintegration, peroxidation of lipids, proteins, and nucleic acids, and the formation of reactive oxygen species (ROS), i.e. superoxide radicals (O_2^-), hydroxyl radicals (OH^-), and H_2O_2. Plants have developd many strategies to cope with stress, including the synthesis of osmoprotectants and enhanced activity of the antioxidative system. In addition, carbohydrate metabolism also appears to play a role in developing tolerance. Several gene and heat shock proteins have also been found to be effective in developing tolerance to stress. Plant growth hormones, abscisic acid, and ethylene have also been identified as regulating plant responses to stress. Other substances, like salicylic acid, jasmonic acid, and NO, also appear to be involved in stress tolerance mechanisms. As such, more research needs to be done on the development of plants with genes that promote the synthesis of these beneficial elements and compounds. Well planned research is needed to investigate the molecular, physiological, and metabolic aspects of abiotic stress tolerance at both the cellular and the whole plant level.

References

Abadi, FSG, Mostafavi, M, Eboutalebi, A, Samavat, S, Ebadi, A. 2010. “Biomass accumulation and proline content of six citrus rootstocks as influenced by long-term salinity”. *Research Journal of Environmental Science*, 4: 158-165.

Ahmad, I, Akhtar, MJ, Zahir, ZA, Jamil, A. 2012. “Effect ofcadmium on seed germination and seedling growth of four wheat (*Triticum aestivum* L.) cultivars”. *Pakistan Journal of Botany*, Vol. 44, No. 5: 1569-1574, 2012.

Ahmed, KA, Tawfik, KM, Zinab, A, El-Gawad, A. 2008. “Tolerance of seven faba bean varieties to drought and salt stress”. *Research Journal of Agricultural and Biological Sciences*, 4: 175-186.

Akhtar, I, Nazir, N. 2013. “Effect of waterlogging and drought stress in plants”. *International Journal of Water Resources and Environmental Sciences*, 2(2): 34-40.

Aksoy, M, & Dinler, BS. 2012. “Changes in physiological parameters and some antioxidant enzymes activities of soybean (*Glycine max* L. Merr.) leaves under cadmium and salt stress”. *Journal of Stress Physiology and Biochemistry*, 8(4): 179-190.

Aldana, F, García, PN, Fischer G. 2014. Effect of waterlogging stress on the growth, development and symptomatology of cape gooseberry

(*Physalis peruviana* L.) plants". *Review Academic Colomb Cienc*, 38(149): 393-400.

Al-Ghamdi, AA. 2009. "Evaluation of oxidative stress tolerance in two wheat (*Triticum aesti- vum*) cultivars in response to drought". *International Journal of Agriculture and Biology*, 11: 7-12.

Amini, F, Ehsanpour, AA. 2005. "Soluble protein, proline, carbohydrates and Na^+/K^+ changes two tomato (*Lycopersicon esculentum* Mill.) cultivars under in vitro salt stress". *American Journal of Biochemistry and Biotechnology*, 1(4): 212-216.

Amirjani, MR. 2010. "Effect of salinity stress on growth, mineral composition, proline content, antioxidant enzymes of soybean". *American Journal of Plant Physiology*, 5: 350-360.

Angelopoulos, K, Dichio, B, Xiloyannis, C. 1996. "Inhibition of photosynthesis in olive trees (Olea europaea L.) during water stress and rewatering". *Journal of Experimental Botany*, 47: 1093-1100.

Araus, JL, Slafer, GA, Reynolds, MP, Royo, C. 2002. "Plant breeding and drought in C3 cereals: what should we breed for?". *Annals of Botany*, 89: 925-940.

Asada K. 1999. "The water-water cycle in chloroplasts: scavenging of active oxygens and dissipation of excess photons". *Annual Review of Plant Biology*, 50: 601-639.

Asada, K. 2006. "Production and scavenging of reactive oxygen species in chloroplasts and their functions". *Plant Physiology*, 141: 391-396.

Ashraf, M, Hafeez, M. 2004. "Thermotolerance of pearl millet and maize at early growth stages: Growth and nutrient relations". *Biologia Plantarum*, 48: 81-86.

Ashraf, M, Akram, NM. 2009. "Improving salinity tolerance of plants through conventional breeding and genetic engineering. An analytical comparison". *Biotechnology Advances*, 27: 744-752.

Ashraf, M. 2009. "Biotechnological approach of improving plant salt tolerance using antioxidants as markers". *Biotechnology Advances*, 27: 84-93.

Ashraf, MA. 2012. "Waterlogging stress in plants. A review". *African Journal of Agricultural Research*, 7(13): 1976-1981.

Azevedo Neto, AD, Nogueira, RJMC, Melo Filho, PA, Santos, R. 2010. "Physiological and biochemical responses of peanut genotypes to water deficit". *Journal of Plant Interactions*, 5: 1-10.

Bargmann, BO, Munnik, T. 2006. "The role of phospholipase D in plantstress responses". *Current Opinion in Plant Biology*, 9: 515-522.

Bavita, A, Shashi, B, Navtej, SB. 2012. “Nitric oxide alleviates oxidative damage induced by high temperature stress in wheat”. *Indian Journal of Experimental Biology*, 50: 372-378.

Beis, A, Patakas, A. 2012. “Relative contribution of photoprotection and anti-oxidative mechanisms to differential drought adaptation ability in grapevines”. *Environmental and Experimental Botany*, 78: 173-183.

Cai, Q, Wang, S, Cui, Z, Sun, J, Ishii, Y. 2004, “Changes in freezing tolerance and its relationship with the contents of carbohydrates and proline in overwintering centipedegrass (*Eremochloa ophiuroides* (Munro) Hack.)”. *Plant Production Science*, 7: 421-426. https://doi.org/10.1626/pps.7.421

Cao, YY, Duan, H, Yang, LN, Wang, ZQ, Liu, L J, Yang, JC. 2009. “Effect of high temperature during heading and early filling on grain yield and physiological characteristics in *indica* rice”. *Acta Agronomica. Sinica*, 35: 512-521.

Cao, YY, Duan, H, Yang, LN, Wang, ZQ, Zhou, SC, Yang JC. 2008. “Effect of heat stress during meiosis on grain yield of rice cultivars differing in heat tolerance and its physiological mechanism”. *Acta Agronomica Sinica*, 34: 2134-2142.

Celik, A, & Atak, C. 2012. “The effect of salt stress on antioxidative enzymes and proline content of two Turkish tobacco varieties”. *Turkish Journal of Biology*, 36: 339-356.

Chakraborty, U, Pradhan, D. 2011. “High temperature-induced oxidative stress in *Lens culinaris*, role of antioxidants and amelioration of stress by chemical pre-treatments”. *Journal of Plant Interaction*, 6: 43-52.

Chibuike GU and Obiora, SC. 2014. “Heavy Metal Polluted Soils: Effect on Plants and Bioremediation Methods”.

Chookhampaeng, S. 2011. “The effect of salt stress on growth, chlorophyll content, proline content and antioxidative enzymes of pepper (*Capsicum annuum* L.) seedling”. *European Journal of Scientific Research*, 49(1): 103-109.

Crafts-Brandner, SJ, Salvucci, ME. 2002. “Sensitivity of photosynthesis in a C_4 plant, maize, to heat stress”. *Plant Physiology*, 129: 1773-1780.

Cramer, GR, Nowak, RS. 1992 “Supplemental manganese improves the relative growth, net assimilation and photosynthetic rates of salt-stressed barley”. *Physiologia Plantarum*. 84: 600-605.

Davies WJ, Zhang J. 1991. “Root signals and the regulation of growth and development ofplants in drying soil”. *Annual Review of Plant Physiology and Plant Molecular Biology*, 42: 55-76.

Diáz, J, Bernal, A, Pomar, F, Merino, F. 2001. “Induction of shikimatedehydrogenase and peroxidase in pepper (*Capsicum annum*

L.) seedlings in response to copper stress and its relation to lignifications". *Plant Science*, 161: 179.

Dionne, J, Castonguay, Y, Nadeau, P, Desjardins, Y. 2001. "Freezing tolerance and carbohydrate changes during cold acclimation of green-type annual bluegrass (*Poa annua* L.) ecotypes". *Crop Science*, 41: 443-451.

Djanaguiraman, M, Prasad, PVV, Al-Khatib, K. 2011. "Ethylene perception inhibitor 1-MCP decreases oxidative damage of leaves through enhanced antioxidant defense mechanisms in soybean plants grown under high temperature stress". *Environmental and Experimental Botany*, 71: 215-223.

Djanaguiraman, M, Prasad, PVV, Seppanen, M. 2010. "Selenium protects sorghum leaves from oxidative damage under high temperature stress by enhancing antioxidant defense system". *Plant Physiology and Biochemistry*, 48: 999-1007.

Djanaguiraman, M, Sheeba, JA, Devi, DD, Bangarusamy, U. 2009. "Cotton leaf senescence can be delayed by nitrophenolate spray through enhanced antioxidant defense system". *Journal of Agronomy and Crop Science*, 195: 213-224.

Djingova, R, Kuleff, I. 2000. "Instrumental techniques for traceanalysis". In *Trace Elements: Their Distribution and Effects in the Environment*, JP Vernet (ed.), Elsevier, London, UK.

Drew, MC, He, CJ., Morgan, PW. 2000. "Programmed cell death and aerenchyma formation in roots". *Trends in Plant Sciences*, 5: 123-127.

Duan, DY, Li, WQ, Liu, XJ, Ouyang, HP. 2007. "Seed germination and seedling growth of *Suaeda salsa* under salt stress". *Annales Botanici Fennici*, 44: 161-169.

Duhan S, Kumari A, Lal, M, Sheokand S. 2019. "Oxidative stress and antioxidant defense under combined waterlogging and salinity stresses" In *Reactive Oxygen, Nitrogen and Sulphur Species in plants*. Edited by Mirza Hasanuzzaman, 113-142 (John Wiley and Sons Ltd).

Duhan, S, Sheokand, S, Kumari, A, Sharma, N. 2017c. "Independent and interactive effects of waterlogging and salinity on carbohydrate metabolism and root anatomy in pigeon pea genotypes at different growth stages". *Indian Journal of Agricultural Research*, 51(3): 197-205.

Duhan, S, Kumari, A, Sheokand, S. 2017a. "Effect of waterlogging and salinity on antioxidative system in pigeon pea plant leaves at different stages of development". *Research on Crops*, 18(3): 559-568.

Duhan, S, Kumari, A, Bala, S, Sharma, N, Sheokand, S. 2018. "Effects of waterlogging, salinity and their combination on stress indices and yield attributes in pigeon pea (*Cajanus cajan* L. Millsp.) genotypes". *Indian Journal of Plant Physiology*, 23(1): 65-76.

Duhan, S, Sheokand S, Kumari, A. 2017d. "Waterlogging, salinity and combined stress a major problem in pigeon pea (*Cajanus cajan* L. Millsp.)". *International Journal of Bioresource and Stress Management*, 8(2): 291-298.

Duhan, S, Sheokand, S. 2020. "Mechanism of Waterlogging Stress Tolerance in Pigeon Pea Plants: Biochemical and Anatomical Adaptation under Waterlogging". In: *Plant Ecophysiology and Adaptation under Climate Change: Mechanism and Perspectives*. Edited by Mirza Hasanuzzaman, 96-105. (Springer Nature Pvt. Ltd).

Duhan, S, Sheokand, S, Kumari, A. 2017b. "Oxidative stress and antioxidative enzymes activity in pigeon pea leaves at different stages of development under waterlogging, salinity and combined stress of waterlogging and salinity". *Journal of Food Legumes*, 30(2): 59-64.

Eckstein, K, Robinson, JC. 1996. "Physiological responses of banana in the subtropics. VI. Seasonal responses of leaf gas exchange to short term water stress". *Journal of Horticultural Science*, 71: 679-692.

Edreira, JIR, Otegui, ME. 2012. "Heat stress in temperate and tropical maize hybrids: Differences in crop growth, biomass partitioning and reserves use". *Field Crops Research*, 130:8 7-98.

Esfandiari, E, Shakiba, MR, Mahboob, SA, Alyari, H, Shahabivand, S. 2008. "The effect of water stress on the antioxidant content, protective enzyme activities, proline content and lipid peroxidation in wheat seedling". *Pakistan Journal of Biological Sciences*, 11: 1916-1922.

Essemine, J, Ammar, S, Bouzid, S. 2010. "Impact of heat stress on germination and growth in higher plants: Physiological, biochemical and molecular repercussions and mechanisms of defence". *Journal of Biological Science*, 10: 565-572.

Evans, DE. 2004. Aerenchyma formation. *New Phytologist*, 161, 35-49.

FAO. 2008. FAO Land and Plant Nutrition Management Service. Available online at http://www.fao. org/ag/agl/agll/spush.

Finkelstein, RR, Gibson, SI. "ABA and sugar interactions regulating development: cross-talk or voices in a crowd?". *Current Opinion in Plant Biology*, 5: 26-32.

Foolad, MR. "Breeding for abiotic stress tolerances in tomato". 2005. In *Abiotic Stresses: Plant Resistance through Breeding and Molecular Approache.* Edited by M Ashraf, PJC Harris, 613-684. (The Haworth Press Inc; New York, USA).

Fortunato, AS, Lidon, FC, Batista-santos, P, Leitào, AE, Pais, IP, Ribeiro, AI, Ramalho, JC. 2010. "Biochemical and molecular characterization of the antioxidative system of Coffea sp. Under cold conditions in genotypes with contrasting tolerance". *Journal of Plant Physiology*, 167: 333-342.

Fowler, S, Thomashow, MF. 2012. "Arabidopsis transcriptome profiling indicates that multiple regulatory pathways are activated during cold acclimation in addition to the CBF cold response pathway". *Plant Cell*, 14: 1675-1690.

Foyer, CH, Noctor, G. 2000. "Oxygen processing in photosynthesis: regulation and signaling". *New Phytologist*, 146(3): 359-388.

Gama, PBS, Inagana, S, Tanaka, K, Nakazawa, R, 2007. "Physiological response of common bean (*Phaseolus Vulg.* L.) Seedlings to salinity stress". *African Journal of Biotechnology*, 2: 79-88.

Ghassemi-Golezani, K, Taifeh-Noori, M, Oustan, SH, Moghaddam, M. 2009. "Response of soybean cultivars to salinity stress". *Journal of Food, Agricuture and Environment*, 7: 401-404.

Greer, DH, Weedon, MM. 2012. "Modelling photosynthetic responses to temperature of grapevine (*Vitis vinifera* cv. Semillon) leaves on vines grown in a hot climate". *Plant Cell Environment*, 35: 1050-1064.

Gull, A, Lone, AA, Wani, NUI. 2019. "Biotic and Abiotic Stresses in Plants". Intech Open.

Gunawardhana, MDM, de Silva, CS. 2011. "Impact of temperature and water stress on growth yield and related biochemical parameters of okra". *Tropical Agricultural Research*, 23: 77-83.

Guo, YP, Zhou, HF., Zhang, L. 2006. "Photosynthetic characteristics and protective mechanisms against photooxidation during high temperature stress in two citrus species". *Scientia Horticulturae*, 108: 260-267.

Gupta, KJ, Stoimenova, M, Kaiser, WM. 2005. "In higher plants, only root mitochondria, but not leaf mitochondria reduce nitrite to NO, *in vitro* and in situ". *Journal of Experimental Botany*, 56: 2601-2609.

Hakeem, KR, Sabir, M, Ozturk, M, Akhtar, MS, Ibrahim, FH. 2017. "Nitrate and nitrogen oxides: sources, health effects and their remediation". In *Reviews of Environmental Contamination and Toxicology*. Edited by F. A. Gunther, P. de Voogt P, 242: 183-217 (Springer, Cham).

Halliwell, B. 2006. "Oxidative stress and neurodegeneration: where are we now?". *Journal of Neurochemistry*, 97: 1634-1658.

Hasanuzzaman, M, Hossain, MA, Fujita, M. 2011. "Nitric oxide modulates antioxidant defense and methylglyoxal detoxification system and reduces salinity-induced damage in wheat seedling". *Plant Biotechnology Reports*, 5: 353-365.

Hasanuzzaman, M, Hossain, MA, Fujita, M. 2012. "Exogenous selenium pretreatmentprotects rapeseed seedlings from cadmium-induced oxidative stress by up-regulating the antioxidant defense and methylglyoxal detoxification systems". *Biological Trace Element Research*

Hasanuzzaman, M, Nahar, K, Fujita, M. 2013. "Extreme Temperature Responses, Oxidative Stress and Antioxidant Defense in Plants". *Intech Open Science*

Hatfield, JL., Boote, KJ, Kimball, BA, Ziska, LH, Izaurralde, RC, Ort, D, Thomson, A, Wolfe, D. 2011. "Climate impacts on agriculture: Implications for crop production". *Agronomy Journal*, 103:351-370.

He, M, He, CQ, Ding, NZ. 2018. "Abiotic Stresses: General Defences of Land Plants and Chances for their Multistress Tolerance". *Frontiers in Plant Scienci.*

Hoffman, L, DaCosta, M, Ebdon JS, Watkins E. 2010. "Physiological changes during cold acclimation of perennial ryegrass accessions differing in freeze tolerance". *Crop Science*, 50: 1037-1047.

Hossain, MA, Araki, H, Takahashi, T. 2011. "Poor grain filling induced by waterlogging is similar to that in abnormal early ripening in wheat in Western Japan". *Field Crops Research*, 123: 100-108.

Hossain, MA, Uddin, SN. 2011: "Mechanisms of waterlogging tolerance in wheat, Morphological and metabolic adaptations under hypoxia or anoxia". *Australian Journal of Crop Science.* 5(9): 1094-1101.

Hu, Y, Jiang, L. Wang, F, Yu, D. 2013. "Jasmonate regulates the inducer of CBF expression-c repeat binding factor/DRE binding factor1 cascade and freezing tolerance in Arabidopsis". *Plant Cell*, 25: 290-292.

Huang, B, Xu, C. 2008. "Identification and characterization of proteins associated with plant tolerance to heat stress". *Journal of Integrated Plant Biology*, 50: 1230-1237.

Huang, JW, Pellet, DM, Papernik, LA, Kochian, LV. 1996. "Aluminum interactions with voltage-dependent calcium transport in plasma membrane vesicles isolated from roots of aluminum-sensitive and resistant wheat cultivars". *Plant Physiology*, 110(2): 561-569.

Huang, M, Guo, Z. 2005. "Responses of antioxidant system to chilling stress in two rice cultivars differing in sensitivity". *Biologia Plantarum*. 49: 81-84.

Hundertmark, Hincha, DK. 2008. "LEA (Late Embryogenesis Abundant) proteins and their encoding genes in *Arabidopsis thaliana*". *BMC Genomics*, 9: 118.

Hurkman, WJ, Vensel, WH, Tanaka, CK, Whitehand, L, Altenbach, SB. 2009. "Effect of high temperature on albumin and globulin accumulation in the endosperm proteome of the developing wheat grain". *Journal of Cereal Science*, 49: 12-23.

Jadia, CD, Fulekar, MH. 2009. "Phytoremediation of heavy metals: recent techniques". *African Journal of Biotechnology*, 8(6): 921-928.

Jaleel, CA, Sankar, B, Sridharan, R, Panneerselvam, R. 2008. "Soil salinity alters growth, chlorophyll content, and secondary metabolite accumulation in *Catharanthus roseus*". *Turkish Journal of Biology*, 32: 79-83.

Jalmi, SK, Bhagat, PK, Verma, D, Noryang, S, Tayyeba, S, Singh, K, Sharma, D, Sinha, AK. 2018. "Traversing the links between heavy metal stress and plant signaling". *Frontiers in Plant Science*, 9: 12.

Jiang, D, Tao, Q, Zhang, G. 2002. "Effect of waterlogging on senescence of flag leaf and root of wheat yangmai". Ying *Yong Sheng Tai Xue Bao*, 13: 1519-1521.

Jithesh, MN, Prashanth, SR, Sivaprakash, KR, & Ajayk, P. 2006. Antioxidative response mechanisms in halophytes, their role in stress defence. *Journal of Genetics*, 85: 237-254.

Karuppanapandian, T, Moon, JC, Kim, C, Manoharan, K, Kim, W. 2011. "Reactive Mechanisms". *Australian Journal of Crop Science*, 5: 709-725.

Kellomäki, S, Strandman, H, Nuutinen, T, Petola, H, Kothonen, KT, Väisänen, H. 2005. "Adaptation of forest ecosystems, Forest and Forestry to Climate Change". FINDAT Working Paper 4. Helsinki: Finnish Environment Institute Mimeographs, 44.

Khan, MA, Weber, DJ. 2008. "Ecophysiology of high salinity tolerant plants (tasks for vegetation science)". (1st edn.) Amsterdam, *Springer*.

Khan, MI, Fatma, M, Per, TS, Anjum, NA, Khan, NA. 2015. "Salicylic acid-induced abiotic stress tolerance and underlying mechanisms in plants". *Frontiers in Plant Science*, 6: 462.

Khan, MN, Mohammad, F, Mobin, M, Saqib, MA. 2014. "Tolerance of plants to abiotic stress: a role of nitric oxide and calcium". In *Nitric oxide in plants: metabolism and role in stress physiology*. Edited by M Khan, M Mobin, F Mohammad, F Corpas F, 225-242 (Springer, Cham, 2014).

Khosravinejab, F, Heydari, R, Farboodnia, T. 2009. "Effect of salinity on organic solutes contents in barley". *Pakistan Journal of Biological Science*, 12: 158-162.

Kibra, MG. 2008. "Effects of mercury on some growth parameters of rice (*Oryza sativa* L.)". *Soil & Environment*, 27(1): 23-28.

Kocsy, G, Szalai, G, Vágújfalvi, A, Stéhli, L, Orosz, G, Galiba, G. 2000. "Genetic study of glutathione accumulation during cold hardening in wheat". *Planta*, 210: 295-301.

Kozlowski, TT, Pallardy, SG. 1997. "Physiology of woody plants". Academic Press, San Diego, CA.

Kramer, U, Cotter-Howells, JD, Charnock, JM, Baker, AJM, Smith, JAC. 1996. "Free histidine as a metal chelator in plants that accumulatenickel". *Nature*, 379(6566): 635-638.

Kuai J, Chen Y, Wang Y, Meng Y, Chen B, Zhao W, Zhou Z. 2016. "Effect of waterlogging oncarbohydrate metabolism and the quality of fiber in cotton (Gossypium hirsutum L.)". *Frontiers in Plant Science*, 7: 877.

Kukreja, S, Nandval, AS, Kumar N, Sharma, SK, Unvi, V, Sharma, PK. 2005. "Plant water status, H_2O_2 scavenging enzymes, ethylene evolution and membrane integrity of *Cicer arietinum* roots as affected by salinity". *Biology of Plants*, 49: 305-308.

Kulz, D. 2003. "Evolution of the cellular stress proteome: From monophyletic origin to ubiquitous function". *Journal of Experimental Biology*, 206: 3119-3124

Kumar, A, Omae, H, Egawa, Y, Kashiwaba, K, Shono, M. 2005. "Proceedings of the International Conference on Sustainable Crop Production in Stress Environment: Management and Genetic Option". JNKVV; Jabarpur, India. "Some Physiological Responses of Snap Bean (*Phseolus vulgaris*L.) to Water Stress during Reproductive Period", 226-227.

Kumar, S, Kaur, R, Kaur, N, Bhandhari, K, Kaushal, N, Gupta, K, Bains, TS, Nayyar, H. 2011. "Heat-stress induced inhibition in growth and chlorosis in mungbean (*Phaseolus aureus* Roxb.) is partly mitigated by ascorbic acid application and is related to reduction in oxidative stress". *Acta Physiolgia Plantarum*, 33: 2091-2101.

Kumutha, D, Ezhilmathi, K, Sairam, RK, Srivastava, GC, Deshmukh, PS, Meena, RC. 2009. "Waterlogging induced oxidative stress and antioxidant activity in pigeon pea genotypes". *Biologia Plantarum*. 53(1): 75-84.

Kurepin, LV, Walton, LJ, Pharis, RP, Emery, RJN, Reid, DM. 2011. "Interactions of temperature and light quality on phytohormone-mediated elongation of *Helianthus annuus* hypocotyls". *Plant Growth Regulation*, 64: 147-154.

Lamattina L, Garcia-Mata C, Graziano M, Pagnussat G 2003. "Nitricoxide: the versatility of an extensive signal molecule". *Annual Review of Plant Biology*, 54: 109-136.

Liu, H, Zhang, J, Christie, P, Zhang, F. 2008. "Influence of iron plaque on uptake and accumulation of Cd by rice (*Oryza sativa* L.) seedlings grown in soil". *Scientia Total Environment*, 394(2-3): 361-368.

Liu, X, Wang, L, Liu, L, Guo, Y, Ren, H. 2011. "Alleviating effect of exogenous nitric oxide in cucumber seedling against chilling stress". *African Journal of Biotechnology*, 10: 4380-4386.

Mahmood, K. 2011. "Salinity tolerance in barley (*Hordeum vulgare* L.). Effects of varying NaCl, K^+ /Na^+ and $NaHCO_3$ levels on cultivars differing in tolerance". *Pakistan Journal of Botany*, 43, 1651-1654.

Majid, M, Ali, A, Essia, B. 2012. "Effect of salinity on sodium and chloride uptake, proline and soluble carbohydrate contents in three alfalfa varieties". *IOSR Journal of Agriculture and Veterinary Science (IOSR-JAVS)*, 1: 1-6.

Marchand, FL, Mertens, S, Kockelbergh, F, Beyens, L, Nijs, I. 2005. "Performance of high arctic tundra plants improved during but deteriorated after exposure to a simulated extreme temperature event". *Global Change in Biology*, 11: 2078-2089.

Marron, N, Maury, S, Rinaldi, C, Brignolas, F. 2006. "Impact of drought and leaf development stage on enzymatic antioxidant system of two Populus deltoids x nigra clones". Annals *of Forest Science*, 63: 323-327.

Meloni, DA, Oliva, MA, Martinez, CA, Cambraia, J. 2003. "Photosynthesis and activity of superoxide dismutase, peroxidase and glutathione reductase in cotton under salt stress". *Environmental and Experimental Botany*, 49: 69-76.

Minglin, L, Yuxiu, Z, Tuanyao, C. 2005. "Identification of genes upregulated in response to Cd exposure in *Brassica juncea* L". *Gene*, 363: 151-158.

Mittler, R. 2002. "Oxidative stress, antioxidants and stress tolerance". *Trends in Plant Science*, 7: 405-410.

Mittler, R. 2017. "ROS are good". *Trends in Plant Science*, 22(1): 11-19.

Mittler, R, Blumwald, E. 2010. "Genetic engineering for modern agriculture: Challenges and perspectives". *Annual Review of Plant Biology*, 61: 443-462.

Mittler, R, Vanderauwera, S, Gollery, M, Van Breusegem, F. 2004. Reactive oxygen gene network of plants. *Trends in Plant Science*, 9: 490-498.

Mohammed, AR, Tarpley, L. 2010. "Effects of high night temperature and spikelet position on yield-related parameters of rice (*Oryza sativa* L.) plants". *European Journal of Agronomy*, 33: 117-123.

Morales, D, Rodriguez, P, Dellamico, J, Nicolas, E, Torrecillas, A, Sanchez-Blanco, MJ. 2003. "High-temperature preconditioning and thermal shock imposition affects water relations, gas exchange and root hydraulic conductivity in tomato". *Biologia Plantatum*, 47: 203-208.

Moreno, AA, Orellana, A. 2011. "The physiological role of the unfolded protein response in plants". *Biological Research*, 44: 75-80.

Moussa, HR, Abdel-Aziz, SM. 2008. "Comparative response of drought tolerant and drought sensitive maize genotypes to water stress". *Australian Journal of Crop Science*, 1: 31-36.

Mukhopadhyay, R, Reddy, PS, Pallela, R, Reddy, NC, Rao, JV, Hash, CT, Kavi Kishore, PB. 2007. "Differential responses of proline, ion accumulation and antioxidative enzyme activities in pearl millet [*Pennisetum glaucum* (L.) R. Br.] lines differing in salt sensitivity". *Journal of Plant Biology*, 34: 185-192.

Munns, R. 1993. "Physiological processes limiting plant growth insaline soils, some dogmas and hypotheses". *Plant Cell and Environment*, 16: 15-24.

Munns, R. 2002: "Comparative physiology of salt and water stress". *Plant Cell and Environment*, 25: 239-250.

Nepomuceno, AL, Oosterhuis, DM, Stewart, JM. 1998. "Physiological responses of cotton leaves and roots to water deficit induced by polyethylene glycol". *Environmental and Experimental Botany*, 40: 29-41.

Noctor, G, Veljovic-Jovanovic, S, Driscoll, S, Novitskaya, L, Foyer, C, 2002. "Drought and oxidative load in the leaves of C_3 plants: A predominant role for photorespiration?". *Annals of Botany*, 89: 841-850.

Ozturk, M, Gucel, S, Sakcali, S, Baslar, S. 2013. "Nitrate and edible plants in the Mediterranean Region of Turkey: an overview". In *Nitrate in Leafy Vegetables-Toxicity and Safety Measures*, edited by S. Umar, 17-51, (IK Intern Publ. House Pvt Ltd, New Delhi).

Pannell, DJ, Ewing, MA. 2004. "Managing secondary dryland salinity: options and challenges". *Agricultural Water Management*, 80: 41-56.

Parida, AK, Das, AB. 2005. "Salt tolerance and salinity effects on plants. A review". *Ecotoxicology and Environmental Safety*, 60: 324-349.

Pinto, AP, Simoes, I, Mota, AM. 2008. "Cadmium impact on root exudates of sorghum and maize plants: a speciation study". *Journal of Plant Nutrition*, 31(10): 1746-1755.

Piramila, BHM, Prabha, AL, Nandagopalan, V, Stanley, AL. 2012. "Effect of heat treatment on germination, seedling growth and some biochemical parameters of dry seeds of black gram". *International Journal of Pharmceutical Phytopharmacology Research*, 1: 194-202.

Pitzschke, A, Forzani, C, Hirt, H. 2006. "Reactive oxygen species signaling in plants". *Antioxid. Redox Signal*, 8: 1757-1764.

Polle, A, Schutzendubel, A. 2004. "Heavy metal signaling in plants: linking cellular and organismic responses". In *Plant responses to abiotic stress*, vol. 4. *Topics in Current Genetics*. Springer, edited by H Hirt, Shinozaki K, 187-215. (Berlin).

Quan, R, Shang, M, Zhang, H, Zhao, Y, Zhang, J. 2004. "Engineering of enhanced glycine betaine synthesis improves drought tolerance in maize". *Plant Biotechnology Journal*, 2: 477-486.

Raven, JA, Evans, MCW, Korb, RE. 1999. "The role of trace metals in photosynthetic electron transport in O_2-evolving organisms". *Photosynthetic. Research*, 60: 111-149.

Reddy, AR, Chaitanya, KV, Jutur, PP, Sumithra, K. 2004. "Differential antioxidative responses to water stress among five mulberry (Morusalba L.) cultivars". *Environmental and Ex- perimental Botany*, 52: 33-42.

Rios, LC, and Pinto, M. 2014. "Effect of salt stress on antioxidant enzymes and lipid peroxidation in leaves in two contrasting corn, 'Lluteno' and 'Jubilee'". *Chilean Journal of Agricultural Research*, 74(1): 89-95.

Robinson, JM, Bunce, JA. 2000. "Influence of drought-induced water stress on soybean and spinach leaf ascorbate-dehydroascorbate level and redox status". *International Journal of Plant Sciences*, 161: 271-279.

Rodríguez, M, Canales, E, Borrás-Hidalgo, O. 2005. "Molecular aspects of abiotic stress in plants". *Biotechnologia Applicada*, 22: 1–10.

Ryu, H, Cho, YG. 2015. "Plant hormones in salt stress tolerance". *Journal of Plant Biology*, 58: 147-155.

Saha, P, Chatterjee, P, Biswas, AK. 2010. "NaCl pretreatment alleviates salt stress by enhancement of antioxidant defense system and osmolyte accumulation in mungbean (*Vigna radiata* L. Wilczek)". *Indian Journal of Experimental Biology*, 48: 593-600.

Sairam, RK, Deshmukh PS, Saxena DC. 1998. "Role of antioxidant systems in wheat genotypes tolerance to water stress". *Biologia Plantarum*, 41(3): 387-394.

Sairam, RK, Kumutha, D, Chinnusamy, V, Meena, RC. 2009. "Waterlogging induced increase in sugar mobilization, fermentation, and related gene expression in the roots of mung bean (*Vigna radiata*)". *Journal of Plant Physiology*, 166: 602-616.

Sairam, RK, Kumutha, D, Ezhilmathi, K, Deshmukh, PS, Srivastava, GC. 2008. "Physiology and biochemistry of waterlogging tolerance in plants". *Biology of Plant*, 52: 401-412.

Sairam, RK, Srivastava, GC, Saxena, DC. 2000. "Increased antioxidant activity under elevated temperature: A mechanism of heat stress tolerance in wheat genotypes". *Biologia Plantarum*, 43: 245-251.

Sanita Di Toppi, L, Gabbrielli, R. 1999. "Response to cadmium in higher plants". *Environmental and Experimental Botany*, 41(2): 105-130.

Sato, S, Kamiyama, M, Iwata, T, Makita, N, Furukawa, H, Ikeda, H. 2006. "Moderate increase of mean daily temperature adversely affects fruit set of *Lycopersicon esculentum* by disrupting specific physiological processes in male reproductive development". *Annals of Botany*, 97: 731-738.

Sayfzadeh, S, Rashidi, M. 2011. "Response of antioxidant enzymes activities of sugar beet to drought stress". *ARPN Journal of Agricultural and Biological Science*, 6(4): 27-33.

Schaller, A, Diez, T. 1991. "Plant specific aspects of heavy metaluptake and comparison with quality standards for food andforage crops". In *Der Einfluß von festen Abfallen auf Boden, Pflanzen*, edited by D. Sauerbeck and S. Lubben, 92-125.

Semenov, MA, Halford, NG. 2009. "Identifying target traits and molecular mechanisms for wheat breeding under a changing climate". *Journal of Experimental Botany*, 60: 2791-2804.

Setter, TL, Waters, I. 2003. "Review of prospects for germplasm improvement for waterlogging tolerance in wheat, barley and oats". *Plant and Soil*, 253: 1-34.

Setter, TL, Waters, I, Sharma, SK, Singh, KN, Kulshreshtha, N, Yaduvanshi, NPS, Ram, PC, Singh, BN, Rane, J, McDonald, G, Khabaz-Saberi, H, Biddulph, TB, Wilson, R, Barclay, I, McLean, R & Cakir, M. 2009. "Review of wheat improvement for waterlogging tolerance in Australia and India, the importance of anaerobiosis and element toxicities associated with different soils". *Annals of Botany*, 103: 221-235.

Sevengor, S, Yasar, F, Kusvuran, S, Ellialtioglu, S. 2011. "The effect of salt stress on growth, chlorophyll content, lipid peroxidation and antioxidative enzymes of pumpkin seedling". *African Journal of Agricultural Research*, 6(21): 4920-4924.

Shao, HB, Chu, LY, Ni, FT, Guo, DG, Li, H, Li, WX. 2010. "Perspective on phytoremediation for improving heavy metal contaminated soils".

In *Plant adaptation and Phytoremediation*, Vol. 2, edited by M. Ashraf, M. Ozturk, M. Ahmad, 227-244.

Sharma, P, Dubey, RS. 2005. "Drought induces oxidative stress and enhances the activities of antioxidant enzymes in growing rice seedlings". *Plant Growth Regulation*, 46: 209-221.

Shi, Q, Ding, F, Wang, X, Wei, M. 2007. "Exogenous nitric oxide protect cucumber roots against oxidative stress induced by salt stress". *Plant Physiology and Biochemistry*, 45(8): 542-550.

Sreeni-vasulu, N, Grimm, B, Wobus, U, Weschke, W. 2000. "Differential response of antioxidant compounds to salinity stress in salt-tolerant and salt-sensitive seedlings of foxtail millet (*Setaria italica*)". *Physiology of Plants*, 109: 435-442.

Srivastava, LM. 2002. "Plant growth and development: hormones and environment". *Annals of Botany*, 92(6): 846, doi:10.1093/aob/mcg209.

Sulaiman, F, Hamzah, R, Asilah, H. 2018. "Heavy metals accumulation in suburban roadside plants of a tropical area (Jengka, Malaysia)" Ecological Processes, 7(1).

Sumesh, KV, Sharma-Natu, P, Ghildiyal, MC. 2008. "Starch synthase activity and heat shock protein in relation to thermal tolerance of developing wheat grains". *Biology of Plants*, 52: 749–753.

Swindell, WR, Huebner, M, Weber, AP. 2007. "Transcriptional profiling of *Arabidopsis* heat shock proteins and transcription factors reveals extensive overlap between heat and non-heat stress response pathways". *BMC Genomics*, 8: 125.

Sytar, O, Kumar, A, Latowski, D, Kuczynska, P, Strzałka, K, Prasad, MNV. 2013. "Heavy metal-induced oxidative damage, defense reactions, and detoxification mechanisms in plants". *Acta Physiologia Plantarum*, 35(4): 985-999.

Taiz, L and Zeiger, E. 2002. *Plant Physiology*. USA: Sinauer Associates, Sunderland.

Takac, T, Luxová, M, Gašparíková, O. 2003. "Cold induced changes in antioxidant enzymes activity in roots and leaves of two maize cultivars". *Biologia Bratislava*, 58: 875-880.

Tanhan, P, Pokethitiyook, P, Chaiyarat, R. 2007. "Uptake and accumulation of cadmium, lead and zinc by Siam weed [Chromolaena odorata (L.)". King & Robinson *Chemosphere*, 68(2): 323-9.

Theocharis, A, Bordiec, S, Fernandez, O, Paquis, S, Dhondt-Cordelier, S, Baillieul, F. 2012a. "*Burkholderia phytofirmans* PsJN primes *Vitis vinifera* L. and confers a better tolerance to low nonfreezing temperatures". *Molecular Plant Microbe Interaction*, 25: 241-249.

Theocharis, AB, Cleament, C, Barka, EA. 2012b. "Physiological and molecular changes in plants grown at low temperatures". *Planta*, 235: 1091-1105.

Thomashow, MF. 1999. "Plant Cold Acclimation: Freezing tolerance genes and regulatory mechanisms". *Ann. Rev. Plant Physiology Plant Molecular Biology*, 50:571-599. https://doi.org/10.1146/annurev.arplant.50.1.571 PMID: 15012220

Tripathy, BC, Oelmuller, R. 2012. "Reactive oxygen species generation and signaling in plants". *Plant Signal Behaviour*, 7(12): 1621-163.

Trischuk, RG, Schilling, BS, Low, NH, Gray, GR, Gusta, LV. 2014. "Cold acclimation, de-acclimation and re-acclimation of spring canola, winter canola and winter wheat: The role of carbohydrates, cold-induced stress proteins and vernalization". *Environmental and Experimental Botany*, 106: 156-163.

Tubiello, FN, Soussana, JF, Howden, SM. 2007. "Crop and pasture response to climate change". *Proc. Nat. Acad. Sci*, USA, 104: 19686-19690.

Turan, O, Ekmekçi, Y. 2011. "Activities of photosystem II and antioxidant enzymes in chickpea (*Cicer arietinum* L.) cultivars exposed to chilling temperatures". *Acta Physiologiae Plantarum*, 33: 67-78.

Unal, T, Aktas, LY, Guven, A. 2014. "Effects of salinity on antioxidant enzymes and proline in leaves of barley seedlings in different growth stages". *Bulgarian Journal of Agricultural Science*, 20 (4), 883-887.

Vasseur, F, Pantin, F, Vile, D. "Changes in light intensity reveal a major role for carbon balance in *Arabidopsis* responses to high temperature". *Plant Cell Environment*, 34: 1563–1576.

Wahid, A. 2007. "Physiological implications of metabolites biosynthesis in net assimilation and heat stress tolerance of sugarcane (*Saccharum officinarum*) sprouts". *Journal of Plant Research*, 120: 219-228.

Wahid, A, Gelani, S, Ashraf, M, Foolad, MR. 2007. "Heat tolerance in plants: An overview". *Environmental and Experimental Botany*, 61: 199-223.

Wang, JZ, Cui, LJ, Wang, Y, Li, JL. 2009. "Growth, lipid peroxidation and photosynthesis in two tall fescue cultivars differing in heat tolerance". *Biologia Plantarum*, 53: 247-242.

Wang, LJ, Li, SH. 2006. "Salicylic acid-induced heat or cold tolerance in relation to Ca^{2+} homeostasis and antioxidant systems in young grape plants". *Plant Science*, 170: 685-694.

Wang, S, Liang, D, Li, C, Hao, Y, Ma, F, Shu, H. 2012. "Influence of drought stress on the cellular ultrastructure and antioxidant system in

leaves of drought-tolerant and droughtsensitive apple rootstocks". *Plant Physiology and Biochemistry*, 51: 81-89.
Wang, W, Vinocur, B, Altman, A. 2003. "Plant responses to drought, salinity and extreme temperatures: towards genetic engineering for stress tolerance". *Planta*, 28: 1-14.
Wang, W, Vinocur, B, Shoseyov, O, Altman, A. 2001. "Biotechnology of plant osmotic stress tolerance: physiological and molecular considerations". *Acta Horticulturae*, 560: 285-292.
Waraich, EA, Ahmad R, Halim A, Aziz T. 2012. "Alleviation of temperature stress by nutrient management in crop plants: a review". *Journal of Soil Science and Plant Nutrition*, 12: 221–244.
Warland, JS, McDonald, MR, McKeown, AM. 2006. "Annual yields of five crops in the family Brassicacae in southern Ontario in relation to weather and climate". *Canadian Journal of Plant Science*, 86: 1209-1215.
Winkel-Shirley, B. 2002. "Biosynthesis of flavonoids and effects of stress". *Current Opinion in Plant Biology*, 5: 218.
Xiaoli, J, Youzong, H, Fanrong, Z, Meixue, Z, & Guoping, Z. 2009. "Genotypic difference in response of peroxidase and superoxide dismutase isozymes and activities to salt stress in barley". *Acta Physiologiae Plantarum*, 31: 1103-1109.
Xin, Z, Browse, J. 2000. "Cold comfort farm: the acclimation of plants to freezing temperatures". *Plant Cell Environment*, 23: 893-902.
Xiumin, C, Yikai, Z, Xiuling, C, Hong, J, Xiaobin, W. 2009. "Effects of exogenous nitric oxide protects tomato plants under copper stress". In *Proceedings of the 3rd international conference on bioinformatics and biomedical engineering (ICBBE'09)*, 1-7.
Xu, S, Li, J, Zhang, X, Wei, H, Cui, L. 2006. "Effects of heat acclimation pretreatment on changes of membrane lipid peroxidation, antioxidant metabolites, and ultrastructure of chloroplasts in two cool-season turfgrass species under heat stress". *Environmental and Experimental Botany*, 56: 274-285.
Yadav S, Atri, N. 2017. "Reactive oxygen species and response of the Calvin–Benson cycle". In *Reactive oxygen species in plants*. 157-163.
Yamamoto, Y, Aminaka, R, Yoshioka, M, Khatoon, M, Komayama, K, Takenaka, D, Yamashita, A, Nijo, N, Inagawa, K, Morita, N. 2008. "Quality control of photosystem II: Impact of light and heat stresses". *Photosynthetic. Research*, 98: 589-608.
Yang, H, Wu, F, Cheng, J. 2011. "Reduced chilling injury in cucumber by nitric oxide and the antioxidant response". *Food Chemistry*, 127: 1237-1242.

Yang, X, Chen, X, Ge, Q, Li, B, Tong, Y, Zhang, A, Li, Z, Kuang, T, Lu, C. 2006. "Tolerance of photosynthesis to photoinhibition, high temperature and drought stress in flag leaves of wheat: A comparison between a hybridization line and its parents grown under field conditions". *Plant Science*, 171: 389-397.

Yin, Y, Li, S, Liao, W, Lu, Q, Wen, X, Lu, C. 2010. "Photosystem II photochemistry, photoinhibition, and the xanthophyll cycle in heat-stressed rice leaves". *Journal of Plant Physiology*, 167: 959-966.

Young, LW, Wilen, RW, Bonham-Smith, PC. 2004. "High temperature stress of *Brassica napus* during flowering reduces micro and megagametophyte fertility, induces fruit abortion, and disrupts seed production". *Journal of Experimental Botany*, 55: 485-495.

Yruela I. 2005. "Copper in plants". *Brazilian Journal of Plant Physiology*, 17(1).

Yuan, HM, Liu, WC, Jin, Y, Lu, YT. 2013. "Role of ROS and auxin in plant response to metal-mediated stress". *Plant Signal Behaviour*, 8(7): e24671.

Zhang, Minfang, Yanming, Ji, Yonghua, Jiang, Zeping, Wang, Lei. 2013. "Effects of salt stress on ion content, antioxidant enzymes and protein profile in different tissues of *Broussonetia papyrifera*". *South African Journal of Botany*, 85: 1-9.

Zhang, X, Cai, J, Wollenweber, B, Liu, F, Dai, T, Cao, W, Jiang, D. 2013. "Multiple heat and drought events affect grain yield and accumulations of high molecular weight glutenin subunits and glutenin macropolymers in wheat". *Journal of Cereal Science*, 57: 134-140.

Zhang, XZ, Ervin, EH, LaBranche, AJ. 2006. "Metabolic defense responses of seeded bermudagrass during acclimation to freezing stress". *Crop Science*, 46:2598-2605.

Zhao, DY, Shen, L, Yu, MM, Zheng, Y, Sheng, JP. 2009. "Relationship between activities of antioxidant enzymes and cold tolerance of postharvest tomato fruits". *Food Science*, 14: 309-313.

Zhao, GQ, Ma, BL, Ren, CZ, 2007. "Growth, gas exchange, chlorophyll fluorescence and ion content of naked oat in response to salinity". *Crop Science*, 47(1): 123-131.

Zheng, C, Jiang, D, Liu, F, Dai, T, Liu, W, Jing, Q, Cao, W. 2009. "Exogenous nitric oxide improves seed germination in wheat against mitochondrial oxidative damage induced by high salinity". *Environmental and Experimental Botany*, 67(1): 222-227.

Zhu, JK. 2001. "Plant salt tolerance". *Trends in Plant Science*, 6: 66-72.

CHAPTER FOUR

THE EFFECTS OF HEAVY METAL POLLUTION ON PLANTS AND THEIR TOLERANCE MECHANISMS

MANJU SHRI AND S. DUBEY

Heavy metal pollution of water and soil resulting in toxicity has become a major issue for crop development and yield. This has been exacerbated by cumulative population growth and increasing food requirements. As with other organisms, plants have acquired a range of mechanisms for the detoxification of heavy metals. These include: signal transduction; adherence of metal compounds to the cell wall and root secretions; metal efflux from the plasma membrane; sequestration within vacuoles by metal binding ligands, such as phytochelatins and metallothioneins; and antioxidant enzymes and heat shock proteins. Substantial progress has been made in recent years to develop transgenic plants for the removal of heavy metals from the environment. This chapter presents an updated understanding of the uptake, accumulation, and detoxification of heavy metals in plants.

Introduction

Over the course of the last few years, the intensification of human activity—rapid industrial development and contemporary agricultural approaches—has resulted in widespread heavy metal pollution, which induces toxicity in living entities (Eapen and D'souza, 2005; Kavamura and Esposito, 2010; Miransari, 2011). Soils have been contaminated with heavy metals through the application of pesticides, fertilizers, and sewage sludge, as well as the poor disposal of heavy metal wastes from smelting and mine tailings (Yang et al., 2005). Though heavy metals are naturally found elements, difficulties emerge when they are discharged in excess into the surroundings due to human activity. Categorised according to their density (>5 g/cm^3), the 53 d-block elements have been characterized as "heavy metals" (Jarup, 2003). Macro and micronutrients play essential roles in fundamental plant processes, such as sugar metabolism, nitrogen fixation, chlorophyll biosynthesis, photosynthesis, DNA synthesis, protein modification, and redox reactions. For instance, Zn is a cofactor for approx. 300 enzymes and 200 transcription factors related to the maintenance of membrane integrity, auxin metabolism, and reproduction (Ricachenevsky et al., 2013).

Nevertheless, at higher concentrations, heavy metals make plants show symptoms of toxicity. As such, their uptake is highly regulated by plant cells (Fidalgo et al., 2013). Some non-essential heavy metals, such as Cd, As, Cr, Pb, Al, and Hg, are toxic even at low concentrations and may inhibit the physiological metabolism of plants (Garzón et al., 2011; Chong-qing et al., 2013). Heavy metals have many detrimental effects directly impacting growth, photosynthesis, chlorosis, the plant water balance, nutrient absorption, senescence, and even triggering plant death.

In addition to their toxic effects on plants, heavy metals also pose a threat to human health due to their persistence in nature. For instance, Pb is a highly toxic heavy metal with a soil retention time of 150-5,000 years and is reported to maintain high concentrations for as long as 150 years (Yang et al., 2005). Plants growing in heavy metal-contaminated sites generally accumulate higher amounts of heavy metals and thus contamination of the food chain occurs. Contaminated food chains are a primary route for the entry of heavy metals into animal and human tissues, causing a number of diseases ranging from dermatitis to various cancers (McLaughlin et al., 1999). This problem can become even worse if sufficient measures are not taken at the right time. Therefore, research in this area is driven by the goal of decreasing the entry of heavy metals into crop plants, thereby reducing the risk of contamination of animals and human beings.

Due to this toxicity, heavy metal contamination in the environment remains a matter of great concern. Several studies reporting the potential of phytoremediation have highlighted the molecular and genetic basis for detoxification of heavy metals in genetically modified plants on heavy metal polluted sites (Yan et al. 2020). This chapter presents current knowledge on heavy metal metabolism, including their uptake, accumulation, and detoxification in plants.

Plant responses to heavy metal stress

As plants are sessile in nature, they cannot move away from environmental stresses. Heavy metal exposure generates wide-ranging physiological and biochemical alterations and plants employ various strategies to survive heavy metal toxicity. Plants counter heavy metal stress through mechanisms that can be characterized as follows: (i) perceiving external stress stimuli; (ii) signal transduction; and (iii) eliciting suitable actions to combat the adverse effects of stress by controlling the physiological, biochemical, and molecular status of cells. It is challenging to identify variations in signal transduction after plants have been subjected to heavy metal stress at the level of the entire plant. However, studying early impacts, for instance oxidative stress, transcriptomic and proteomic alterations, and metabolite accumulation, can help us examine signal transduction changes when plants are exposed to stress. For example, overexpression of genes related to dehydration stress in barley after exposure to Cd and Hg was reported by Tamás et al. (2010). In the same way, during heavy metal exposure, oxidative stress and glutathione reduction can be used as early signals of sensing and signal transduction in alfalfa roots, as reported by Hernandez et al. (2012). Zhang et al. (2002) studied how seed germination and seedling growth in wheat was suppressed when exposed to high arsenic concentrations. Likewise, a decline in plumule and radicle length of *Helainthus annuus L.* plantlets after As exposure was reported by Imran et al. (2013). Furthermore, it has been found that arsenic causes a decline in the photosynthetic pigment, impairs the chloroplast membrane, and reduces enzyme activity by binding with the sulfhydryl group of proteins; it has also been reported that it modifies the nutrient balance and protein metabolism (Ahsan et al., 2010). It has been suggested that heavy metals cause toxicity in plants through four mechanisms. These consist of: (i) resemblance with nutrient cations, which creates competition for entry at the root surface; (ii) binding of heavy metals to the sulfhydryl group (SH), leading to the disruption of the structure and function of proteins, which renders them inactive; (iii) dislocation of vital cations from precise binding sites leading to reduced activity; and (iv) overproduction of reactive oxygen species (ROS), which subsequently impair important biomolecules (Sharma and Dietz, 2009). To examine the effect of this stressor, plant roots, the first point of contact with heavy metals, have been extensively used. Plants growing in heavy metal polluted areas show reduced growth, performance, and yield (Keunen et al., 2011; Anjum et al., 2014). Root development relies on cell division and elongation and inhibited root growth due to a reduction in the mitotic process has been found in numerous plants upon heavy metal exposure (Thounaojam et al., 2012). A higher toxic effect on cell division during Cr (VI) stress in comparison to Cr (III) treatment was reported by Liu et al. (1992), while Sundaramoorthy et al. (2010) reported that Cr (VI) triggered a delay in the cell cycle that tended to decrease root growth due to the suppression of cell division. Pena et al. (2012) examined how Cd toxicity and ROS production alters the G1/S transition and progression through the S phase by inhibited cyclin dependent kinase (CDK) expression. It has been observed that higher concentrations of Cu influence both elongation and meristem zones by modifying auxin supply through the PINFORMED1 (PIN1) protein (Yuan et al., 2013). Likewise, Petö et al. (2011) confirmed that superfluous Cu hinders root length and modifies root structure by causing changes in auxin concentration, which disturbs nitric oxide function. The plant cytoskeleton is also a target of heavy metal stress, as shown by a reduction in root growth with an increase in root diameter (Zobel et al., 2007). As such, much research has proposed that heavy metals produce a reduction in root growth that disturbs water and nutrient uptake, thus disturbing their movement to the above-ground plant parts, adversely influencing shoot development, and lessening biomass accumulation. Roots apply numerous approaches to address these issues. These include the production and deposition of callose to reduce or avoid heavy metal toxicity by constraining the entry of heavy metals and intensifying the elasticity of the root framework. To improve plant tolerance to HM stress, roots also permit their uptake and transport to above ground plant parts for vacuole compartmentalization (Fahr et al., 2013). Plasma membranes function as an extremely restrictive entry check point for undesirable elements into cells. Kenderešová et al. (2012) examined the greater resistence of *Arabidopsis halleri* and *Arabidopsis arenosa* to metal toxicity in comparison to *Arabidopsis thaliana* on account of their lower membrane depolarization, signifying that fluctuations in the membrane's electric potential would be a good means of checking the effects of heavy metal toxicity. After entering the cell, heavy metals modify metabolic activities, inducing decreased growth and a decline in biomass cumulation (Nagajyoti et al., 2010). Tinted stem and root length, as well as chlorosis in younger leaves, may result due to metal stress and this can spread to mature leaves after extended exposure (Srivastava et al., 2012). Heavy metals are extremely toxic and affect the cellular and molecular mechanism of plants, causing modifications in their physiological and biochemical processes. They can negatively impact seed growth, photosynthesis, gas exchange, and respiration; denature enzymes; inactivate important molecules by blocking their functional groups; and alter hormone proportions, nutrient absorption, and protein and DNA synthesis (Singh et al., 2013). Their toxic effects have been investigated in relation to several photosynthetic indicators, such as photosynthetic rate (Pn) and intracellular CO_2 concentration (Ci) in tomato plantlets under Cd exposure (Dong et al., 2005). Substantial decay in chlorophyll levels, followed by a decrease in the photochemical proficiency of photosystem II (PSII), was confirmed by Maleva et al. (2012) with Mn, Cu, Cd, Zn, and Ni stress in *Elodea Densa*. Li et al. (2012) investigated the reduced chlorophyll and carotenoids levels and the quantum yield of PSII in *Thalassia hemprichii* under Cu, Zn, Pb, and Cd exposure, suggesting that photosynthesis is highly affected by metal toxicity. In addition, heavy metals reduce CO_2 absorption by either lessening the activity of RUBP carboxylase, or by binding with the thiol group of RUBISCO. For example, Monnet et al. (2001) reported that Zn impeded the activity of RUBISCO in *Phaseolusvulgaris* by exchanging Zn^{+2} for Mg^{+2}. Decreased RUBISCO activity was also observed by Muthuchelian et al. (2001) during Cd stress in *Erythrina variegate*. These scientists also found diminished CO_2 fixation, which may be due to a reduction in ATP (Husaini and Rai, 1991), as Cd ions decrease the proton source for reduction reactions (Ferretti et al., 1993). Likewise, it was found that Cu decreases RUBISCO activity in *Chenopodium rubrum* (Schafer et al., 1992) by reacting with the essential cysteine residue of the enzyme (Siborova, 1988). This reduction in gas exchange, pigments, photosynthetic level, quantum yield of PSII, stomatal conductivity, and CO_2 absorption is possibly due to structural alterations prompted by metal toxicity. The impacts emerging as a result of such structural changes have been stated by numerous research studies (Sánchez-Pardo et al., 2012). Furthermore, heavy metals also exhibit destructive consequences on other physiological practices, such as nitrogen metabolism, thus interfering with plant growth. Increased protease activity has been reported to be a consequence of heavy metal toxicity (Chaffei et al., 2003), minimizing the function of enzymes associated with nitrate and ammonia assimilation. It has been shown that nitrogen assimilation is affected by Cd stress in terms of nitrate uptake and transportation, nitrate reductase, and GS activity inhibition (Lea and Miflin, 2004). Heavy metal-facilitated modification of hormonal proportions is related to their cytotoxicity in plants (Wilkinson et al., 2012). As an example, remarkable alterations have been detected in auxin intensity—indole-3-acetic acid (IAA), indole-3-butyricacid (IBA), and naphthalene acetic acid (NAA)—and in the expression of about 69 microRNAs in *Brassica juncea* under As stress (Srivastava et al., 2013). Conversely, *B. juncea* growth was improved with additional delivery of

IAA during As exposure, suggesting an approach to counterbalance the impact of As stress by regulation of the hormone level.

Heavy metal uptake and transport

Hyperaccumulator plants have a phenomenal capability to take up heavy metals from the soil (Yang et al., 2002). However, the uptake of heavy metals by hyperaccumulators is influenced by numerous factors, such as water content, pH, and organic constituents. Likewise, metals need an appropriate transportation mode to enter the plant. A number of investigators have studied how pH influences proton excretion by roots, which acidify the rhizosphere, hence improving metal disintegration, as well as the development of metal-accumulator plants (Kuriakose and Prasad, 2008). Organic materials discharged from the roots affect the growth of hyperaccumulator plants. Krishnamurti et al. (1997) found that Cd solubility was affected by organic acids released via Cd complex formation. Heavy metal mobilization and enhanced absorption results due to the release of pH and organic substances from the rhizosphere of the hyperaccumulating plant (Peng et al., 2005). Elevated heavy metal uptake has also been connected to boosted root propagation (Whiting et al., 2000) and has been attributed to the upregulation of certain genes. To pinpoint the genes involved in overexpression, numerous experiments have been undertaken using the hyperaccumulator species *Arabidopsis halleri* and *Thlapsi caerulescens* and congener non-hyperaccumulating species. Studies on *T. caerulescens* and *A. halleri* have revealed that increased Zn uptake is due to overexpression of genes belonging to the ZIP (zinc/iron-regulated transporter-like proteins) family encoding plasma membrane-located transporters (Assunção et al., 2001). These are ZTN1 and ZTN2 in *T. caerulescens* and ZIP6 and ZIP9 in *A. halleri*. The decreased uptake of Cd at increasing Zn concentrations has been noted in both genera. This clearly demonstrates that expression of the ZIP genes is Zn regulated (Assunção et al., 2010) and Cd influx is mainly due to Zn transporters having a strong preference for Zn over Cd (Weber et al., 2006). Evidence exists that As, being a chemical analog of phosphate, enters the plant cell via phosphate transporters (Kanoun-Boulé et al., 2009). Similarly, a study of the As hyperaccumulator *Pteris vittata* and non-hyperaccumulator *Pteris tremula* showed that the plasma membranes in root cells of *P. vittata* had a higher density of phosphate/arsenate transporters than *P. tremula* (Caille et al., 2005), which is possibly due to constitutive gene overexpression. In addition, a study into the Se hyperaccumulators *Astragalus bisulcatus* (Fabaceae) and *Stanleya pinnata* (Brassicaceae) revealed that there was a higher Se/S ratio in the shoots of the Se hyperaccumulator species in comparison to the non-hyperaccumulator sister species. This observation is also supported by the fact that enhanced Se uptake used sulfate transporters (Galeas et al., 2007).

Root-to-shoot translocation of heavy metals

In contrast to non-hyperaccumulator plants, hyperaccumulators do not retain heavy metals absorbed through the roots but translocate them from root to shoot efficiently via the xylem. There are various proteins that help in this translocation: heavy metal-transporting ATPases (CPx-type, P1B-type); cation diffusion facilitator (CDF) proteins; natural resistance-associated macrophage proteins (Nramp); MATE (multidrug and toxin efflux) proteins; and zinc–iron permease (ZIP) proteins. Different proteins have roles in the uptake of various heavy metals. The CPx-type ATPases are involved in transporting toxic metals like Cd, Cu, Zn, and Pb using ATP across cell membranes (Williams et al., 2000). The P1B-type ATPases have a comparable role transporting heavy metals, but they regulate metal homeostasis as well as tolerance (Axelsen and Palmgren, 1998). Studies have shown the upregulation of these heavy metal ATPases in hyperaccumulator plants in comparison to non-accumulator plants, suggesting their importance for hyperaccumulation (Papoyan and Kochian, 2004). Along with these HMAs, another class of proteins, Nramp, has also been shown to be involved in transporting heavy metal ions. In rice, three Nramps—OsNramp1, OsNramp2, and OsNramp3—have been found to be expressed in different tissues and transport distinct, but related ions (Belouchi et al., 1997). Cation diffusion facilitators (CDF) have been found to be involved in the transportation of Zn, Co, and Cd, as well as regulating the efflux of cations out from the cytoplasmic compartment (Mäser et al., 2001). ZNT1, a member of the ZIP family, has been found to be expressed at high levels in the roots and shoots of *T. caerulescens* (Pence et al., 2000). In the root tissues of Arabidopsis the another ZN transporter ZAT1 was found to be expressed at higher level. (Van der Zaal et al., 1999). Members of the MATE family has also been found to be involved in the transport of heavy metals: FDR3, a MATE protein, was found to be expressed in the roots tissues of *T. caerulescens* and *A. halleri* (Talke et al., 2006). The above mentioned studies all provide strong evidence that multiple transporter proteins are involved in the translocation of heavy metals.

Different strategies used by plants to detoxify heavy metals

1. Heavy metals and signal transduction in plants

To deal with heavy metal stress, different strategies are used by plants, including compartmentalization and metal export among others. When a plant senses increased heavy metal concentrations in its vicinity, a complicated signal transduction network is activated. Stress signaling molecules and stress-related proteins are synthesized. As a result, specific genes are transcriptionally activated that are specific to metal stress, hence the response to heavy metals is epigenetically controlled (Cicatelli et al., 2013). This mechanism requires the coordination of complex biochemical and physiological processes. Generally, the signaling pathway consists of a sequential strategy: perception of a heavy metal concentration; activation of signaling molecules like lipids; and the modulation of endocytic pathways (Galvan-Ampudia and Testerink, 2011). Along with plant developmental processes, stress responses are also mediated by other signaling molecules like jasmonates (lipid derived signals) and annexins, etc. (Jami et al., 2010). Different heavy metals induce different responses with different signaling molecule cascades, but a general signal transduction pathway involves the following: ROS; signaling pathway; calcium-calmodulin system; phosphorylation cascade; mitogen-activated protein (MAP) kinase; and certain hormones (DalCorso et al. 2010).

2. Phosphorylation cascades

In many cellular processes, phosphorylation is a major event (Chen et al. 2013). Under stress, different proteins of the thylakoid membranes undergo phosphorylation or dephosphorylation in response to biotic or abiotic stresses (Tikkanen and Aro, 2012). In most cases, phosphorylation occurs on threonine and serine residues, but it can also occur on tyrosine residues resulting in many developmental and stress responses (Sasabe et al., 2011). Alteration of cytokinesis involves the activation of MAPKKK and proteins like mitotic kinesin. This activation is done by cyclin dependent kinases (Sasabe et al., 2011).

3. MAP kinase

Downstream signaling events after sensing ROS in plant cells include calmodulin (the Ca-binding protein) and the activation of phospholipid signaling and G-proteins, ultimately leading to the accretion of phosphatidic acid and/or activation of MAPK pathways (De Pinto et al. 2012). In this signaling mechanism, the pathway has three kinases that are activated sequentially: MAPK kinase kinase (MAPKK); MAPK kinase (MAPKK or MKK); and MAP kinases (MAPK or MPK) (Opdenakker et al., 2012). These kinases then phosphorylate different cellular compartments. This signaling pathway has been shown to be initiated as a result of different stresses (DalCorso et al., 2010). In reaction to a type of stress, e.g. cold in plants, the MAP kinase cascade is activated. In turn, this activates other signaling pathways, resulting in an adaptative response to the stress (Yang et al., 2010). Thanks to this pathway, as the concentration of stress causing agents increases, the plant's growth, cell division, and differentiation decrease (Smekalova et al., 2013). A study showed that MPK3 and MPK6 are activated in *A. thaliana* in response to short term exposure to $CdCl_2$ at low concentrations (1 μM), following the accumulation of ROS (Liu et al., 2010). In another study, a rice MAPK cascade called Oryza sativa MAPKK (OSMKK4 and OSMKK3) was analyzed. When exposed to arsenite, rice plants showed elevated transcription levels of OSMKK3 in leaves and roots after 30 minutes, while levels of OSMKK4 were raised after 3 hours (Rao et al., 2011). This cascade plays diverse roles in signaling and transferring information from sensors to responders and controlling processes like proliferation, differentiation, and death. Different biotic and abiotic stresses produce different responses; these stresses include temperature extremes, heavy metals, salinity, high/low osmolarity, and drought etc. (Pitzschke and Hirt, 2010).

4. Calcium-calmodulin system

This system uses proteins as secondary messengers for plant responses to the presence of heavy metals. Under stress conditions, the concentration of calcium greatly increases in the cell, stimulating calmodulin proteins. This system regulates a variety of mechanisms including the regulation of certain genes, transport of ions across the membranes, metabolic processes, and tolerance of different metals (Dal Corso et al., 2010). This system is primarily activated in response to stress and has a significant role in cold tolerance and acclimatization in coordination with other signaling pathways (Yang et al., 2010). It has been found that some heavy metals, such as Cd and Cu, disturb intracellular levels of Ca, impairing the calmodulin signaling pathway, which regulates Cd tolerance mechanisms. It was found that transgenic tobacco plants overexpressing calmodulin were effectively able to resist Ni and Pb toxicity (Lindberg et al. 2012). A regulated member of this family, CRLK1 (calcium-calmodulin receptor like kinase 1), enhances cold tolerance in plants even under freezing conditions (Yang et al., 2010a).

5. The role of hormones in heavy metal stress

During heavy metal signaling, particular hormones have a very important role in the regulation of plant development, growth, and reproduction, as well as key functions in the regulation of defense mechanisms against biotic and abiotic stresses (Pieterse et al., 2012). This regulation is in response to concentrations of different metals (DalCorso et al., 2010). Recently, plant hormones such as indole-3-acetic acid (IAA), also known as auxins, brassinosteroids (BR), and abscisic acid (ABA), have been found to play a vital role in stress management (Vázquez et al., 2013). Auxins seem to play an important role in stress regulation as shown by exogenous ABA inducing overproduction of phytochelatins (PC) under Cd, Zn, and Cu stress in *Prosopis juliflora* (Usha et al., 2009). Steroid hormones, also known as brassinosteroids (BR), are critical to developmental and detoxification processes—they promote the formation of PCs, which chelate heavy metal ions accumulated in plant cells (Choudhary et al., 2010). In another study, the effects of exogenous application of BRs on raddish and mustard plants under copper stress were studied. It was found that BR promoted shoot and root growth by overcoming copper toxicity—BRs were found to be responsible for antioxidant activity and increased PC formation (Choudhary et al., 2011a). When combined, secondary messengers, like hormones and MAPK, ensure effective transcription and signaling under stress conditions (Smekalova et al., 2013). Ethylene is a hormone normally produced in response to age-related stresses (Khan et al., 2014). Jasmonic acid and salicylic acid are also produced. The hormone signaling pathway has positive and negative regulators that are crucial to hormonal crosstalk in stress and defense mechanisms. Auxins, cytokinins, ABA, and brassinosteroids, etc. adapt to changes and outcomes in antagonistic and synergistic connections that play important roles in abiotic stress tolerances (Ha et al., 2012). In a recent study, various concentrations of EBL (24-epibrassinolide), a brassinosteroid obtained from *Brassica juncea*, were applied to seeds of the same species 8 hours prior to sowing. These seeds were then exposed to Ni stress and showed improved growth with lower uptake of Ni ions (Kanwar et al., 2013). The use of an EBL foliar spray also improved the antioxidant system by promoting the formation of catalase, superoxide dismutase, proline, and peroxidase in bean plants with Cd toxicity (Rady, 2011). Choudhary et al. (2011b) studied the influence of EBL on other hormones and found that, under Cr stress, EBL increased the formation of IAA to promote seedling growth in raddish plants. It was also seen to enhance the production of ABA, increasing Cr tolerance (Choudhary et al., 2011b).

6. Formation of metal complexes

Another important strategy for heavy metal detoxification is heavy metal chelation by high affinity ligands. In plants, two types of metal binding peptides are produced—phytochelatins (PC) and metallothioneins (MT). Besides these, many other small molecules are also used in metal chelation inside the cytosol.

7. Phytochelatins

Phytochelatins (PC) are a family of polypeptides that are rich in cysteine. PCs are synthesized from GSH (glutathione) and are found in plants and fungi. The enzyme, PC synthase, is activated when metal ions bind to it, converting GSH to PC. Heavy metals ions, such as Hg, Cu, Ni, Au, Zn, and Cd, induce the biosynthesis of PCs; Cd is known to be the strongest PC inducer. The PC-Cd complex is formed by the attachment of Cd to PC through the thiolic group (-SH) of the cysteine residue. These complexes are sequestered into the vacuole by ABC proteins, as mentioned above. Plants are unable to metabolize or neutralize Cd. Rather, they limit Cd circulation in the cytosol or transport it away through the xylem and phloem. PCs also contribute to homeostasis of Cu and Zn by providing transitory storage for the ions. It has been found that both heavy metal resistant and heavy metal susceptible plants produce PCs. The process of detoxification is not limited to metal chelation: after the chelated metal complex is transported into the vacuole, it is stabilized there by the formation of complexes with organic acids or sulfides. As part of this complex mechanism, PCs also transport metal ions.

8. Metallothioneins

Like PCs, metallothioneins (MT) are a major family of cysteine rich, metal binding low molecular weight peptides found in many organisms. Although the structure of plant MTs is different to those found in other organisms, they interact with heavy metals through the thiolic group of their cysteine residues. It has been suggested that they are involved in homeostasis and the sequestering of important heavy metal ions (Coldsbrough, 2010). They also provide protection against intracellular oxidative damage. Based on the arrangement of cysteine residues, MTs can be divided into three classes. Different isoforms of MTs exist and their capability to bind and confiscate heavy metals also varies. Hormones, cytotoxic agents, and heavy metal ions, such as Cu, Au, Zn, Hg, Ni, Co, and Cd, are MT inducers. A study showed three B. rapa MT genes (BrMT1-BrMT3) to be differentially regulated in various heavy metal stresses and the expression of all 3 genes was variable when the seedlings were treated with Fe. Upon Cu exposure, BrMT1 expression was increased compared to BrMT2, whereas BrMT3 expression remained unchanged (Ahn et al., 2012). The tonoplast ABC transporter protein was also up-regulated in transgenic OsMT1 plants, which sequestered Cd metal ions inside the vacuole, thus assisting Cd detoxification (Yang and Chu, 2011). Although animal and fungal MTs have a clear role in heavy metal detoxification, the precise relationship between plant MTs and heavy metals remains unclear.

9. Amino acids, organic acids, and phosphate derivatives

Organic acids, including malate, citrate, and oxalate, have the ability to bind metals and are therefore deployed in heavy metal tolerance mechanisms. Organic acids confer metal tolerance in two ways—external exclusion and internal tolerance. In the former, organic acids are secreted from plant roots and make a stable metal-ligand complex with metal, thus hindering metal ions from entering and accumulating in sensitive sites in the roots (Sharma and Dietz, 2009). Citrate is synthesized in plants by the citrate synthase enzyme, which has higher affinity for metals compared to malate and oxalate. It has been found that citrate plays a principal role in Fe chelation, although other heavy metal ions, like Zn, Ni, Cd, and Co, also have high affinity for citrate. The amount of citrate produced depends greatly on Ni exposure (Hassan and Aarts, 2011).

Amino acids and derivatives chelate metals thus conferring plant resistance against toxic metals. Histidine is considered the most important free amino acid in heavy metal metabolism. The presence of putative carboxyl, amino, and imidazole groups makes histidine a versatile metal chelator (Krämer, 2010). It has been found that histidine plays a major role in tolerance to nickel. In Ni hyperaccumulator species *Alyssum* and *N. goesingense*, the concentration of histidine in the xylem exudate is higher compared to closely related nonaccumulator species (McNear et al., 2010). Nicotianamine (NA) is a non-proteogenic and low molecular weight amino acid that is found in root and leaf cells and also in the phloem (Hassan and Aarts, 2011). It is synthesized as a result of a condensation reaction of three S-adenosyl-Lmethionine molecules catalyzed by NA synthase (Talke et al., 2006). NA chelates Cu, Fe, and Zn by complex formation, before storing them in the vacuole. In *A. thaliana* and other plants, NA is involved in the influx and efflux of Cu, Zn, and Fe by transporting metals from one cell to another (Klatte et al., 2009). The membrane YSL transporter family transports the metal-NA complexes; such complexes are substrates for these transporters (Gendre et al., 2007). Phytate is a principal form of stored phosphorus in plant cells. The molecule consists of six phosphate groups, which allow the chelation of various cations, including Ca, Mg, Fe, Mn, and K (Kumar et al., 2010).

10. Oxidative stress defense and damaged proteins repair mechanisms

If the intracellular concentration of metal ions overwhelms the plant's defense mechanisms and strategies, it begins to suffer oxidative stress due to the inhibition of metal dependent antioxidant enzyme and the production of methylglyoxal (MG) and reactive oxygen species (ROS) (Hossain and Fujita, 2010). This includes the induction of enzymes such as catalase (CAT) and superoxide dismutase (SOD), and the production of non-enzymatic free radical scavengers. The literature shows many cases of such induction. For example, APX and CAT production were found to be induced in response to excess Fe exposure in the leaves of *Nicotiana plumbaginifolia*. Similarly, CAT3 was induced in *B. juncea* plants in response to Cd exposure (Minglin et al., 2005). Cd has been show to cause oxidation of CAT in pea plants, reducing plant activity. The plant responds by upregulating the transcription of the CAT gene (Romero-Puertas et al., 2007). SOD activity was shown to increase in response to prolonged activity of metals and wheat leaves were shown to respond to excess Cd by increasing SOD levels (Lin et al., 2007).

11. Glutathione (GSH)

Production of ROS also encourages the activation of the ascorbic acid glutathione scavenging system. GSH is a low molecular weight, non-enzymatic antioxidant; it is a major redox buffer and antioxidant found abundantly in all plant cell compartments (Yadav, 2010). GSH plays a role in the control of H_2O_2 levels and up-regulation of GSH is of crucial importance because it induces defensive strategies against ROS and MG through various pathways, including the activation and expression of enzymes associated with GSH and argininosuccinate (AsA) (Hossain et al., 2012). GSH plays a key role in responding to various environmental stresses, metal tolerance, and metal chelation because it acts as an ROS scavenger and a substrate for PC biosynthesis (Krämer, 2010). GSH shields proteins against denaturation triggered by oxidation under stress conditions. It is also indirectly involved in protecting membranes by maintaining a reduced state of zeaxanthin and α-tocopherol (Hossain and Fujita, 2011). GSH (Glu-Cys-Gly) is the major intracellular antioxidant inside cells and is a precursor to PCs; it also forms complexes with heavy metals, such as Cd (Wójcik and Tukiendorf, 2011). Nevertheless, the role of GSH as an antioxidant depends on its intracellular concentration and varies substantially under Cd toxicity. It has been observed in relation to *B. juncea* and *B. campestris* that an increase in Cd concentration increases the GSH concentration in response (Anjum et al., 2008). In *Phaseolus vulagaris* and *Pisum sativum*, Cd treatment induces ascorbate peroxidase (APX) (Romero-Puertas et al., 2007). An elevation in GSH concentration in the leaves, roots, and stems under various metal stresses (Hg, Cd, and Pb) was reported by Huang et al. (2010). In *Holcus lanatus*, an As tolerant species, GSH levels increased significantly, induced by As, compared to species that were As-sensitive (Verbruggen et al., 2009). Recent studies have demonstrated that nitric oxide (NO) influences GSH synthesis, as revealed in a study with *Medicago truncatula* in which increased NO in the roots increased GSH and glutathione (GS) gene expression (Xu et al., 2011). S-nitrosoglutathione (GSNO) is formed during the interaction of GSH with NO, possibly interconnecting the reactive nitrogen and ROS based signaling pathways (Xiong et al., 2010).

12. Glutathione S-transferases (GST)

GSTs belong to a superfamily of multifunctional phase II metabolic isoenzymes, which are best known for their ability to detoxify xenobiotics. GSTs catalyze the conjugation of a reduced form of GSH with various compounds to form derivatives that can be sequestered in the vacuole or secreted from the cell. In addition, they also defend against oxidants and abiotic induced oxidative stress (Hossain et al., 2012b). Increased GST activity was seen in barley (*Hordeum vulgare*) when subjected to Cu, Hg, Co, Cd, Pb, and Zn (Valentovičová et al., 2009). Another study on rice seedlings showed increased activity of GST in response to Cd stress (50 μM Cd, 7 days) (Hu et al., 2009). GST activity was shown to increase substantially in the callus of onions in response to Cd stress (1 mM CdCl 2) (Rohman et al., 2010). Time and dose affects GST levels in plants and thus it is likely that, in the case of severe toxicity and when basal antioxidant mechanisms are used up and depleted, more effective responses like GST are activated, (Hossain et al., 2010). Astudy by Dixit et al. (2011) observed that tobacco plants with upregulated GST gene expression show less lipid peroxidation due to reduced Cd accumulation than wild type plants, indicating better Cd tolerance.

13. Glutathione peroxidases (GPX)

GPXs belong to family of enzymes with peroxidase activity. They are an important part of the antioxidant network of plants and found in different cellular organelles. Their primary activity is the catalyzation of the reduction process of free H_2O_2 to water, ROOHs (organic hydroperoxidases), and lipid hydroperoxidases to alcohol in the presence of GSH and other reducing agents (Foyer and Noctor, 2011). Mammalian glutathione peroxidases possess higher affinity to lipid peroxidases than H_2O_2 and are homologous to most of the plant's GPX genes. *A. thaliana* exposed to Cd stress (11 and 10 μM for 7 days) showed an increase in GPX activity, indicating protection by GPX against lipid peroxidases (Semane et al., 2007). Additionally, it was found that Cu, Hg, and Ni induced increased GPX activity, whereas Co, Zn, and Pb exposure had no substantial effect on activity (Valentovičová et al., 2009). It was reported that GPX activity is induced under As stress in other plants (Gupta et al., 2009).

14. Dehydroascorbate reductase (DHAR)

DHAR is an important enzyme required in the AsA-GSH (ascorbate-glutathione) reaction in higher plants. Ascorbic acid is oxidized through spontaneous disproportion to form dehydroascorbic acid (DHA). DHAR then reduces DHA to AsA using GSH (Yang et al., 2009). It was observed that the enzyme activity of DHA reductase increased in two barley genotypes when exposed to Cd stress for 1-25 days (Chen et al., 2010). In the same way, DHA reductase activity was amplified in wheat leaves and roots when exposed to different concentrations of Cd (Paradiso et al., 2008). Under Ni stress, its activity was shown to increase in rice roots and shoots, showing that Ni maintains elevated levels of AsA by activating the AsA regenerating system (Maheshwari and Dubey, 2009). An increase in DHAR activity was seen in two tomato cultivars, "Josefina" and "Kosaco", when subjected to boron stress, in addition to a reduction in AsA content, indicating that enzyme activity increases to reduce oxidative stress (Cervilla et al., 2007). When exposed to Cd stress transgenic, *A. thaliana* plants, in which genes for CAT and GST were overexpressed, saw a considerable increase in DHA reductase activity was seen; however, it was reduced in non-transgenic plants (Zhao et al., 2009).

15. Glutathione reductase (GR)

GR belongs to an enzyme family that catalyzes the reduction of GSSG to GSG using NADPH. Heavy metals, in particular Cd, reduce the GSH/GSSG ratio and activate antioxidant enzymes, such as GR and SOD. The cysthiol group of reduced glutathione (GSH) is oxidized and glutathione reductase (GR) catalyzes the reverse reaction using NADPH, thus acting as a defense mechanism against Cd-generated oxidative stress (Yannarelli et al., 2007). GR helps cells to resist toxicity caused by RO metabolites. It maintains the reduced form of GSH and ascorbate in cells, which sequentially retain the cellular redox state of heavy metal stress (Hossain et al., 2011). In their study, Nouairi et al. (2009) reported significantly increased glutathione reductase activity in *B. napus* at lower concentrations of Cd ions, which dropped when concentrations were raised after 15 days of treatment. In *B. juncea* leaves, GR enzyme activity levels were unaffected by variable concentrations of Cd ions. In a study on two cultivars of mung beans, GR activity was increased in both Cd tolerant and Cd sensitive genotypes in response to Cd stress (Anjum et al., 2011). However, in another study it was observed that GR activity decreases considerably in response to Cd stress on mung bean seedlings (Hossain et al., 2010), indicating that GR activity is greatly influenced by genotypic differences. Another study reported that GR enzymatic activity in *B. napus* roots decreased, while in leaves the glutathione reductase buildup was comparatively high when exposed to Cu (Russo et al., 2008). Upregulation of GR contributes to the maintenance of the GSH/GSSG ratio and higher GSH levels under heavy metal stress, which are, in turn, used by many enzymes that depend on GSH involved in ROS and MG metabolism (Hossain et al., 2012).

16. Heat shock proteins (HSP)

Heat shock proteins are signaling molecules released in response to metal induced and other forms of abiotic stress. HSPs are found in all types of cell and are not just expressed in response to elevated temperatures, but in response to other stresses as well (Dubey, 2011). They protect and repair proteins and act as molecular chaperons to ensure correct folding. The induction of HSPs by several heavy metal ions (Al, Cu, Hg, Cd, and Zn) has been reported (Dubey, 2011).

Conclusion

With the growth in the human population, different human activities have resulted in the increase of toxic compounds in soil that are not naturally present. Heavy metal pollution brings various health problems and risks and can result in accumulation of heavy metals in the food chain of both plants and animals. Plants have developed certain defense mechanisms to shield themselves and respond to stimuli to minimize the risks. Antioxidant systems, both enzymatic and nonenzymatic, symptoms, and tolerance mechanisms vary from plant to plant. This chapter has explained the different mechanisms used by plants to defend themselves against various toxic heavy metals.

References

Ahn, Young Ock, Sun Ha Kim, Jeongyeo Lee, HyeRan Kim, Haeng-Soon Lee, and Sang-Soo Kwak. 2012. "Three *Brassica rapa* metallothionein genes are differentially regulated under various stress conditions". *Molecular biology reports* 39, No. 3: 2059-2067.

Ahsan, Nagib, Dong-Gi Lee, Kyung-Hee Kim, Iftekhar Alam, Sang-Hoon Lee, Ki-Won Lee, Hyoshin Lee, and Byung-Hyun Lee. 2010. "Analysis of arsenic stress-induced differentially expressed proteins in rice leaves by two-dimensional gel electrophoresis coupled with mass spectrometry". *Chemosphere* 78, No. 3: 224-231.

Anjum, Naser, A, Sarvajeet, S, Gill, Ritu Gill, Mirza Hasanuzzaman, Armando C Duarte, Eduarda Pereira, Iqbal Ahmad, Renu Tuteja, and Narendra Tuteja. 2014. "Metal/metalloid stress tolerance in plants: role of ascorbate, its redox couple, and associated enzymes". *Protoplasma*, 251, No. 6: 1265-1283.

Anjum, Naser Aziz, Shahid Umar, Altaf Ahmad, Muhammad Iqbal, and Nafees Ahmad Khan. 2011. "Cadmium causes oxidative stress in moongbean [*Vigna radiata* (L.) Wilczek] by affecting antioxidant enzyme systems and ascorbate-glutathione cycle metabolism". *Russ. J. Plant Physiol.*, 58: 92-99.

Anjum, Naser Aziz, Shahid Umar, Altaf Ahmad, Muhammad Iqbal, and Nafees Ahmad Khan. 2008. "Ontogenic variation in response of *Brassica campestris* L. to cadmium toxicity". *J. Plant Interact.*, No. 3: 189-198.

Assunção, AG, Da Costa Martins, P, De Folter, S, Vooijs, R, Schat, H, and Aarts, MGM. 2001. "Elevated expression of metal transporter genes in three accessions of the metal hyperaccumulator Thlaspi caerulescens". *Plant Cell Environ.*, 24, No. 2: 217-226.

Assunção, Ana GL, Eva Herrero, Ya-Fen Lin, Bruno Huettel, Sangita Talukdar, Cezary Smaczniak, Richard GH Immink et al. 2010. "*Arabidopsis thaliana* transcription factors bZIP19 and bZIP23 regulate the adaptation to zinc deficiency". *Proceedings of the National Academy of Sciences*, 107, No. 22: 10296-10301.

Axelsen, Kristian B, and Michael G Palmgren. 1998. "Evolution of substrate specificities in the P-type ATPase superfamily". *Journal of molecular evolution*, 46, No. 1: 84-101.

Belouchi, Abdelmajid, Tony Kwan, and Philippe Gros. 1997. "Cloning and characterization of the OsNramp family from *Oryza sativa*, a new family of membrane proteins possibly implicated in the transport of metal ions". *Plant Mol. Biol.* 33, No. 6: 1085-1092.

Caille, N, Zhao, FJ, and McGrath, SP. 2005. *New Phytol.* 165: 755-761.

Chen, Fei, Fang Wang, Feibo Wu, Weihua Mao, Guoping Zhang, and Meixue Zhou. 2010. "Modulation of exogenous glutathione in antioxidant defense system against Cd stress in the two barley genotypes differing in Cd tolerance". *Plant Phys. Biochem.* 48, No. 8: 663-672.

Chen, Yang-Er, Zhong-Yi Zhao, Huai-Yu Zhang, Xian-Yin Zeng, and Shu Yuan. 2013. "The significance of CP29 reversible phosphorylation in thylakoids of higher plants under environmental stresses". *J. Exp. Bot.* 64, No. 5: 1167-1178.

Chiraz, Chaffei, Gouia Houda, and Ghorbel Mohamed Habib. 2003. "Nitrogen metabolism in tomato plants under cadmium stress". *J. Plant Nutr.* 26, No. 8: 1617-1634.

Choudhary, Sikander Pal, Mukesh Kanwar, Renu Bhardwaj, BD Gupta, and RK Gupta. 2011b. "Epibrassinolide ameliorates Cr (VI) stress via influencing the levels of indole-3-acetic acid, abscisic acid, polyamines and antioxidant system of radish seedlings". *Chemosphere* 84, No. 5: 592-600.

Choudhary, Sikander Pal, Renu Bhardwaj, BD Gupta, Prabhu Dutt, RK Gupta, Mukesh Kanwar, and Stefania Biondi. 2011a. "Enhancing effects of 24-epibrassinolide and Putrescine on the antioxidant capacity and free radical scavenging activity of *Raphanus sativus* seedlings under Cu ion stress". *Acta Phys Plant* 33, No. 4: 1319-1333.

Choudhary, Sikander Pal, Renu Bhardwaj, Bishan Datt Gupta, Prabhu Dutt, Rajinder Kumar Gupta, Stefania Biondi, and Mukesh Kanwar. 2010. "Epibrassinolide induces changes in indole-3- acetic acid, abscisic acid and polyamine concentrations and enhances antioxidant potential of radish seedlings under copper stress". *Phys. Plant.* 140, No. 3: 280-296.

Cicatelli, Angela, Valeria Todeschini, Guido Lingua, Stefania Biondi, Patrizia Torrigiani, and Stefano Castiglione. 2013. "Epigenetic control of heavy metal stress response in mycorrhizal versus non-mycorrhizal poplar plants". *Environmental Science and Pollution Research* 21, No. 3: 1723-1737.

DalCorso, Giovanni, Silvia Farinati, and Antonella Furini. 2010. "Regulatory networks of cadmium stress in plants". *Plant Signaling & Behavior* 5, No. 6: 663-667.

De Pinto, MC, V Locato, and L De Gara. 2012. "Redox regulation in plant programmed cell death". *Plant, cell & environment* 35, No. 2: 234-244.

Dubey R. 2011. "Metal toxicity, oxidative stress and antioxidative defense system in plants". In Gupta SD (ed.) *Reactive Oxygen Species and Antioxidants in Higher Plants*. CRC Press, Boca Raton, 177-203.

Eapen, Susan, and SF D'souza. 2005. "Prospects of genetic engineering of plants for phytoremediation of toxic metal". *Biotechnology advances* 23, No. 2: 97-114.

Fahr, Mouna, Laurent Laplaze, Najib Bendaou, Valerie Hocher, Mohamed El Mzibri, Didier Bogusz, and Abdelaziz Smouni. 2013. "Effect of lead on root growth". *Front. Plant Sci.* 4: 175.

Ferretti, M, R Ghisi, L Merlo, F Dalla Vecchia, and C Passera. 1993. "Effect of cadmium on photosynthesis and enzymes of photosynthetic sulfate and nitrate assimilation pathways in maize (*Zea mays* L.)". *Photosynthetica* 29: 49-54.

Fidalgo, Fernanda, Manuel Azenha, António F Silva, Alexandra de Sousa, Ana Santiago, Pedro Ferraz, and Jorge Teixeira. 2013. "Copper-induced stress in *Solanum nigrum* L. and antioxidant defense system responses". *Food and Energy Security* 2, No. 1: 70-80.

Foyer, Christine, H, and Graham Noctor. 2011. "Ascorbate and glutathione: the heart of the redox hub". *Plant physiology* 155, No. 1: 2-18.

Galeas, Miriam, L, Li Hong Zhang, John, L, Freeman, Mellissa Wegner, and Elizabeth AH Pilon-Smits. 2007. "Seasonal fluctuations of selenium and sulphur accumulation in selenium hyperaccumulators and related nonaccumulators". *New Phytologist* 173, No. 3: 517-525.

Galvan-Ampudia, Carlos S., and Christa Testerink C. 2011. "Salt stress signals shape the plant root". *Current Opinion in Plant Biology* 14, No. 3: 296-302.

Garzón Garzón, Teresa, Benet Gunsé, Ana Rodrigo Moreno, A. Deri Tomos, Juan Barceló, and Charlotte Poschenrieder. 2011. "Aluminium-induced alteration of ion homeostasis in root tip vacuoles of two maize varieties differing in Al tolerance". *Plant Science* 180, No. 5: 709-715.

Gendre, Delphine, Pierre Czernic, Geneviève Conéjéro, Katia Pianelli, Jean-François Briat, Michel Lebrun, and Stéphane Mari. 2007. "TcYSL3, a member of the YSL gene family from the hyperaccumulator Thlaspi caerulescens, encodes a nicotianamine-Ni/Fe transporter". *The Plant Journal* 49, No. 1: 1-15.

Goldsbrough, Peter. 2010. "12 Metal tolerance in plants: The role of phytochelatins and methallothioneins". *Phytorem. Contam. Soil Water*, 221.

Gupta, Meetu, Pallavi Sharma, Neera Bhalla Sarin, and Alok Krishna Sinha. 2009. "Differential response of arsenic stress in two varieties of *Brassica juncea* L.". *Chemosphere* 74, No. 9: 1201-1208.

Ha, Sukbong, Radomira Vankova, Kazuko Yamaguchi-Shinozaki, Kazuo Shinozaki, and Lam-Son Phan Tran. 2012. "Cytokinins: metabolism and function in plant adaptation to environmental stresses". *Trends in Plant Science* 17, No. 3: 172-179.

Halušková, L'ubica, Katarína Valentovičová, Jana Huttová, Igor Mistrík, and Ladislav Tamás L. 2009. "Effect of abiotic stresses on glutathione peroxidase and glutathione Stransferase activity in barley root tips". *Plant Physiology and Biochemistry* 47, No. 11-12: 1069-1074.

Hassan, Zeshan, and Mark GM Aarts. 2011. "Opportunities and feasibilities for biotechnological improvement of Zn, Cd or Ni tolerance and accumulation in plants". *Environmental and Experimental Botany* 72, No. 1: 53-63.

Hernández, Luis, E, Cristina Ortega-Villasante, M Belén Montero-Palmero, Carolina Escobar, and Ramón O Carpena. 2012. "Heavy metal perception in a microscale environment: a model system using high doses of pollutants". In *Metal Toxicity in Plants: Perception, Signaling and Remediation*, 23-39.

Hossain, MA, and M Fujita. 2011. "Regulatory role of components of ascorbate-glutathione (AsA-GSH) pathway in plant tolerance to oxidative stress". In Anjum NA, Umar S, Ahmad A (eds.), *Oxidative Stress in Plants: Causes, Consequences and Tolerance*, 81-147.

Hossain, Mohammad Anwar, Mirza Hasanuzzaman, and Masayuki Fujita. 2010. "Up-regulation of antioxidant and glyoxalase systems by exogenous glycinebetaine and proline in mung bean confer tolerance to cadmium stress". *Molecular Biology of Plants* 16, No. 3: 259-272.

Hossain, Mohammad Anwar, Pukclai Piyatida, Jaime A. Teixeira da Silva, and Masayuki Fujita. 2012. "Molecular mechanism of heavy metal toxicity and tolerance in plants: central role of glutathione in detoxification of reactive oxygen species and methylglyoxal and in heavy metal chelation". *Journal of Botany*.

Huang, Guo-Yong, You-Shao Wang, Cui-Ci Sun, Jun-De Dong, and Zong-Xun Sun. 2010. "The effect of multiple heavy metals on ascorbate, glutathione and related enzymes in two mangrove plant seedlings (*Kandelia candel* and *Bruguiera gymnorrhiza*)". *Ocean Hydrobiol. Studies* 39, No. 1: 11-25.

Hussain, Y, and RC Rai. 1991. "Studies on nitrogen and phosphorus metabolism and the photosynthetic electron transport system of Nostoc linckia under cadmium stress". *J. Plant Physiol.* 138: 429-435.

Imran, Muhammad Asif, Rass Masood Khan, Zulfiqar Ali, and Tariq Mahmood. 2013. "Toxicity of arsenic (As) on seed germination of sunflower (*Helianthus annuus* L.)". *International Journal of Physical Sciences* 8, No. 17: 840-847.

Jami, Sravan Kumar, Robert D Hill, and PB Kirti. 2010. "Transcriptional regulation of annexins in Indian Mustard, Brassica juncea and detoxification of ROS in transgenic tobacco plants constitutively expressing AnnBj1". *Plant Signaling & Behavior* 5, No. 5: 618-621.

Järup, Lars. 2003. "Hazards of heavy metal contamination". *British Medical Bulletin* 68, No. 1: 167-182.

Jing, Dong, Wu, Feibo, and Zhang Guoping. 2005. "Effect of cadmium on growth and photosynthesis of tomato seedlings". *J. of Zhejiang University Science B* 6, No. 10: 974-980.

Kanoun-Boulé, Myriam, Joaquim AF Vicente, Cristina Nabais, MNV Prasad, and Helena Freitas. 2009. "Ecophysiological tolerance of duckweeds exposed to copper". *Aquatic Toxicology* 91, No. 1: 1-9.

Kanwar, Mukesh Kumar, Renu Bhardwaj, Sikandar Pal Chowdhary, Priya Arora, Priyanka Sharma, and Subodh Kumar. 2013. "Isolation and characterization of 24-Epibrassinolide from *Brassica juncea* L. and its effects on growth, Ni ion uptake, antioxidant defense of Brassica plants and in vitro cytotoxicity". *Acta Physiologiae Plantarum* 35, No. 4: 1351-1362.

Kavamura, Vanessa Nessner, and Elisa Esposito. 2010. "Biotechnological strategies applied to the decontamination of soils polluted with heavy metal". *Biotechnology Advances* 28, no. 1: 61-69.

Kenderešová, Lucia, Andrea Staňová, Ján Pavlovkin, Eva Ďurišová, Miriam Nadubinská, Milada Čiamporová, and Miroslav Ovečka. 2012. "Early Zn^{2+}-induced effects on membrane potential account for primary heavy metal susceptibility in tolerant and sensitive Arabidopsis species". *Annals of Botany* 110, No. 2: 445-459.

Keunen, Els, Tony Remans, Sacha Bohler, Jaco Vangronsveld, and Ann Cuypers. 2011. "Metal-induced oxidative stress and plant mitochondria". *International Journal of Molecular Sciences* 12, No. 10: 6894-6918.

Khan, Mamoona, Wilfried Rozhon, and Brigitte Poppenberger. 2014. "The role of hormones in the aging of plants: a mini-review". *Gerontology* 60, No. 1: 49-55.

Klatte, Marco, Mara Schuler, Markus Wirtz, Claudia Fink-Straube, Rüdiger Hell, and Petra Bauer. 2009. "The analysis of Arabidopsis nicotianamine synthase mutants reveals functions for nicotianamine in seed iron loading and iron deficiency responses". *Plant Physiology* 150, No. 1: 257-271.

Krämer, Ute. 2010. "Metal hyperaccumulation in plants". *Annual Review of Plant Biology* 61: 517-534.

Krishnamurti, GSR, G Cieslinski, PM Huang, and KCJ Van Rees. 199. "Kinetics of cadmium release from soils as influenced by organic acids: implication in cadmium availability". *Journal of Environmental Quality* 26, No. 1: 271-277.

Kumar, Vikas, Amit K Sinha, Harinder PS Makkar, and Klaus Becker. 2010. "Dietary roles of phytate and phytase in human nutrition: a review". *Food Chemistry* 120, No. 4: 945-959.

Kuriakose, Saritha V, and MNV Prasad. 2008. "Cadmium stress affects seed germination and seedling growth in *Sorghum bicolor* (L.) Moench by changing the activities of hydrolyzing enzymes". *Plant growth regulation* 54, No. 2:143-156.

Lea, Peter J, and Ben J Miflin. 2004. "Glutamate synthase and the synthesis of glutamate in plants". *Plant Physiology and Biochemistry* 41, No. 6-7: 555-564.

Li, Lei, Xiaoping Huang, Devajit Borthakur, and Hui Ni. 2012. "Photosynthetic activity and antioxidative response of seagrass *Thalassia hemprichii* to trace metal stress". *Acta Oceanologica Sinica* 31, No. 3: 98-108.

Lin, Fanyun, Jianhong Xu, Jianrong Shi, Hongwei Li, and Bin Li. 2010. "Molecular cloning and characterization of a novel glyoxalase I gene TaGly I in wheat (*Triticum aestivum* L.)". *Molecular Biology Reports* 37, No. 2: 729-735.

Lindberg, Sylvia, Md Abdul Kader, and Vladislav Yemelyanov. 2012. "Calcium signalling in plant cells under environmental stress". In *Environmental Adaptations and Stress Tolerance of Plants in the Era of Climate Change*, 325-360. Springer, New York, NY.

Liu, Donghua, Wusheng Jiang, and Maoxie Li. 1992. "Effects of trivalent and hexavalent chromium on root growth and cell division of Allium cepa". *Hereditas* 117, No. 1: 23-29.

Liu, Xiao-Min, Kyung Eun Kim, Kang-Chang Kim, Xuan Canh Nguyen, Hay Ju Han, Mi Soon Jung, Ho Soo Kim et al. 2010. "Cadmium activates Arabidopsis MPK3 and MPK6 via accumulation of reactive oxygen species". *Phytochemistry* 71, No. 5-6: 614-618.

Maheshwari, Ruchi, and RS Dubey. 2009. "Nickel-induced oxidative stress and the role of antioxidant defence in rice seedlings". *Plant Growth Regulation* 59, No. 1: 37-49.

Maleva, MG, GF Nekrasova, GG Borisova, NV Chukina, and OS Ushakova. 2012. "Effect of heavy metal on photosynthetic apparatus and antioxidant status of elodea". *Russian Journal of Plant Physiology* 59, No. 2: 190-197.

Mäser, Pascal, Sébastien Thomine, Julian I Schroeder, John M Ward, Kendal Hirschi, Heven Sze, Ina N Talke et al. 2001. "Phylogenetic relationships within cation transporter families of Arabidopsis". *Plant Physiology* 126, No. 4: 1646-1667.

McLaughlin, MJ, Parker, DR, and Clark, JM. 1999. "Metal and micronutrients-food safety issues". *Field Crops Research* 60, No. 1-2: 143-163.

McNear Jr, David H, Rufus L Chaney, and Donald L Sparks. 2010. "The hyperaccumulator Alyssum murale uses complexation with nitrogen and oxygen donor ligands for Ni transport and storage". *Phytochemistry* 71, No. 2-3: 188-200.

Minglin, Lang, Zhang Yuxiu, and Chai Tuanyao. 2005. "Identification of genes upregulated in response to Cd exposure in *Brassica juncea* L.". *Gene*. 363: 151-158.

Miransari, Mohammad. 2011. "Hyperaccumulators, arbuscular mycorrhizal fungi and stress of heavy metal". *Biotechnology Advances* 29, No. 6: 645-653.

Monnet, Fabien, Nathalie Vaillant, Philippe Vernay, Alain Coudret, Huguette Sallanon, and Adnane Hitmi. 2001. "Relationship between PSII activity, CO_2 fixation, and Zn, Mn and Mg contents of Lolium perenne under zinc stress". *Journal of Plant Physiology* 158, No. 9: 1137-1144.

Muthuchelian, Krishnasamy, Massimo Bertamini, and Namachevayam Nedunchezhian. 2001. "Triacontanol can protect Erythrina variegata from cadmium toxicity". *Journal of Plant Physiology* 158, No. 11: 1487-1490.

Nagajyoti, PC, KD Lee, and TVM Sreekanth. 2010. "Heavy metal, occurrence and toxicity for plants: a review". *Environmental Chemistry Letters* 8, No. 3: 199-216.

Nouairi, Issam, Wided Ben Ammar, Nabil Ben Youssef, Douja Daoud Ben Miled, Mohamed Habib Ghorbal, and Mokhtar Zarrouk. 2009. "Antioxidant defense system in leaves of Indian mustard (Brassica juncea) and rape (*Brassica napus*) under cadmium stress". *Acta Physiologiae Plantarum* 31, No. 2: 237-247.

Opdenakker, Kelly, Tony Remans, Jaco Vangronsveld, and Ann Cuypers. 2012. "Mitogen-activated protein (MAP) kinases in plant metal stress: regulation and responses in comparison to other biotic and abiotic stresses". *International Journal of Molecular Sciences* 13, No. 6: 7828-7853.

Papoyan, Ashot, and Leon V Kochian. 2004. "Identification of Thlaspi caerulescens genes that may be involved in heavy metal hyperaccumulation and tolerance. Characterization of a novel heavy metal transporting ATPase". *Plant Physiology* 136, No. 3: 3814-3823.

Paradiso, Annalisa, Rosalia Berardino, Maria C de Pinto, Luigi Sanità di Toppi, Maria M Storelli, Franca Tommasi, and Laura De Gara. 2008. "Increase in ascorbate–glutathione metabolism as local and precocious systemic responses induced by cadmium in durum wheat plants". *Plant and Cell Physiology* 49, No. 3: 362-374.

Pena, Liliana B, Roberto A Barcia, Claudia E Azpilicueta, Andrea AE Méndez, and Susana M Gallego. 2012. "Oxidative post translational modifications of proteins related to cell cycle are involved in cadmium toxicity in wheat seedlings". *Plant Science* 196: 1-7.

Pence, Nicole S, Paul B Larsen, Stephen D Ebbs, Deborah LD Letham, Mitch M Lasat, David F Garvin, David Eide, and Leon V Kochian. 2000. "The molecular basis for heavy metal hyperaccumulation in *Thlaspi caerulescens*". *Proceedings of the National Academy of Sciences* 97, No. 9: 4956-4960.

Peng, Hong-Yun, Xiao-E Yang, Li-Yang Jiang, and Zhen-Li He. 2005. "Copper phytoavailability and uptake by *Elsholtzi asplendens* from contaminated soil as affected by soil amendments". *Journal of Environmental Science and Health* 40, No. 4: 839-856.

Pető, Andrea, Nóra Lehotai, Jorge Lozano-Juste, José León, Irma Tari, László Erdei, and Zsuzsanna Kolbert. 2011. "Involvement of nitric oxide and auxin in signal transduction of copper-induced morphological responses in Arabidopsis seedlings". *Annals of Botany* 108, No. 3: 449-457.

Pieterse, Corné MJ, Dieuwertje Van der Does, Christos Zamioudis, Antonio Leon-Reyes, and Saskia CM Van Wees. 2012. "Hormonal modulation of plant immunity". *Annual Review of Cell and Developmental Biology* 28: 489-521.

Pitzschke, Andrea, and Heribert Hirt. 2010. "Mechanism of MAPK-targeted gene expression unraveled in plants". *Cell Cycle* 9: 18-19.

Rady, Mostafa M. 2011. "Effect of 24-epibrassinolide on growth, yield, antioxidant system and cadmium content of bean (*Phaseolus vulgaris* L.) plants under salinity and cadmium stress". *Scientia Horticulturae* 129, No. 2. 129: 232-237.

Rao, Kudupudi Prabhakara, Gubbala Vani, Kundan Kumar, Dhammaprakash Pandhari Wankhede, Mohit Misra, Meetu Gupta, and Alok Krishna Sinha. 2011. "Arsenic stress activates MAP kinase in rice roots and leaves". *Archives of Biochemistry and Biophysics* 506, No. 1: 73-82.

Ricachenevsky, Felipe Klein, Paloma Koprovski Menguer, Raul Antonio Sperotto, Lorraine Elizabeth Williams, and Janette Palma Fett. 2013. "Roles of plant metal tolerance proteins (MTP) in metal storage and potential use in biofortification strategies". *Frontiers in Plant Science* 4: 144.

Rohman, MM, MS Uddin, and M Fujita. 2010. "Up-regulation of onion bulb glutathione S-transferases (GSTs) by abiotic stresses: a comparative study between two differently sensitive GSTs to their physiological inhibitors". *Plant Omics* 3, No. 1: 28-34.

Romero-Puertas, María C, Francisco J Corpas, María Rodríguez-Serrano, Manuel Gómez, A. Luis, and Luisa M. Sandalio. 2007. "Differential expression and regulation of antioxidative enzymes by cadmium in pea plants". *Journal of Plant Physiology* 164, No. 10: 1346-1357.

Russo, M, Cristina Sgherri, Riccardo Izzo, and F Navari-Izzo. 2008. "*Brassica napus* subjected to copper excess: Phospholipases C and D and glutathione system in signaling". *Environmental and Experimental Botany* 62, No. 3: 238-246.

Sánchez-Pardo, Beatriz, Mercedes Fernández-Pascual, and Pilar Zornoza. 2012. "Copper microlocalisation, ultrastructural alterations and antioxidant responses in the nodules of white lupin and soybeanplants grown under conditions of copper excess". *Environmental and Experimental Botany* 84: 52-60.

Sasabe, Michiko, Véronique Boudolf, Lieven De Veylder, Dirk Inzé, Pascal Genschik, and Yasunori Machida. 2011. "Phosphorylation of a mitotic kinesin-like protein and a MAPKKK by cyclin-dependent kinases (CDKs) is involved in the transition to cytokinesis in plants". *Proceedings of the National Academy of Sciences* 108, No. 43: 17844-17849.

Schäfer, CH Simper, and B Hofmann. 1992. "Glucose feeding results in coordinated changes of chlorophyll content, ribulose-1,5-bisphosphate carboxylase-oxygenase activity and photosynthetic potential photoautotrophic suspension cultured cells of Chenopodium ruburum". *Plant, Cell & Environment* 15(3): 343-350.

Semane, Brahim, Ann Cuypers, Karen Smeets, Frank Van Belleghem, Nele Horemans, Henk Schat, and Jaco Vangronsveld. 2007. "Cadmium responses in *Arabidopsis thaliana*: glutathione metabolism and antioxidative defence system". *Physiologia Plantarum* 129, No. 3: 519-528.

Sharma, Shanti S., and Karl-Josef Dietz. 2009. "The relationship between metal toxicity and cellular redox imbalance". *Trends in Plant Science* 14, No. 1: 43-50.

Singh, Vijay Pratap, Prabhat Kumar Srivastava, and Sheo Mohan Prasad. 2013. "Nitric oxide alleviates arsenic-induced toxic effects in ridged Luffa seedlings". *Plant Physiology and Biochemistry* 71: 155-163.

Smékalová, Veronika, Anna Doskočilová, George Komis, and Jozef Šamaj. 2013. "Crosstalk between secondary messengers, hormones and MAPK modules during abiotic stress signalling in plants". *Biotechnology Advances* 32, No. 1: 2-11.

Srivastava, Garima, Sushil Kumar, Gunjan Dubey, Vagish Mishra, and Sheo Mohan Prasad. 2012. "Nickel and ultraviolet-B stresses induce differential growth and photosynthetic responses in *Pisum sativum* L. seedlings". *Biological Trace Element Research* 149, No. 1: 86-96.

Srivastava, Sudhakar, Ashish Kumar Srivastava, Penna Suprasanna, and SF D'souza. 2013. "Identification and profiling of arsenic stress-induced microRNAs in *Brassica juncea*". *Journal of Experimental Botany* 64, No. 1: 303-15.

Stiborova, Marie. 1988. "Cd2+ ions affect the quaternary structure of ribulose-1,5-bisphosphate carboxylase from barley leaves". *Biochemie und Physiologie der Pflanzen* 183, No. 5: 371378.

Sundaramoorthy, Perumal, Alagappan Chidambaram, Kaliyaperumal Sankar Ganesh, Pachikkaran Unnikannan, and Logalakshmanan Baskaran. 2010. "Chromium stress in paddy: (i) nutrient status of paddy under chromium stress; (ii) phytoremediation of chromium by aquatic and terrestrial weeds". *Comptes rendus biologies* 333, No. 8: 597-607.

Talke, Ina N, Marc Hanikenne, and Ute Krämer. 2006. "Zinc-dependent global transcriptional control, transcriptional deregulation, and higher gene copy number for genes in metal homeostasis of the hyperaccumulator *Arabidopsis halleri*". *Plant Physiology* 142, No. 1: 148-167.

Tamás, Ladislav, Igor Mistrík, Jana Huttová, L'ubica Halušková, Katarína Valentovičová, and Veronika Zelinová. 2010. "Role of reactive oxygen species-generating enzymes and hydrogen peroxide during cadmium, mercury and osmotic stresses in barley root tip". *Planta* 231, No. 2: 221-231.

Thounaojam, Thorny Chanu, Piyalee Panda, Purabi Mazumdar, Devanand Kumar, Gauri Dutta Sharma, Lingaraj Sahoo, and Panda Sanjib. 2012. "Excess copper induced oxidative stress and response of antioxidants in rice". *Plant Physiology and Biochemistry*, 53: 33-39.

Tikkanen, Mikko, and Eva-Mari Aro. 2012. "Thylakoid protein phosphorylation in dynamic regulation of photosystem II in higher plants". *Biochimica et Biophysica Acta (BBA)-Bioenergetics* 1817, No. 1: 232-238.

Usha, B, Gayatri Venkataraman, and Ajay Parida. 2009. "Heavy metal and abiotic stress inducible metallothionein isoforms from *Prosopis juliflora* (SW) DC show differences in binding to heavy metals in vitro". *Molecular Genetics and Genomics* 281, No. 1: 99-108.

Van der Zaal, Bert J, Leon W Neuteboom, Johan E Pinas, Agnes N Chardonnens, Henk Schat, Jos AC Verkleij, and Paul JJ Hooykaas. 1999. "Overexpression of a novel Arabidopsis gene related to putative zinc-transporter genes from animals can lead to enhanced zinc resistance and accumulation". *Plant Physiology* 119, No. 3: 1047-1055.

Vázquez, Miriam Núñez, Yanelis Reyes Guerrero, Lisbel Martínez González, and Walfredo Torres de la Noval. 2013. "Brassinosteroids and Plant Responses to Heavy Metal Stress. An Overview". *Open J. Metal* 3, 3: 34.

Verbruggen, Nathalie, Christian Hermans, and Henk Schat. 2009. "Mechanisms to cope with arsenic or cadmium excess in plants". *Current Opinion in Plant Biology* 12, No. 3: 364-372.

Wang, Chong-qing, Wang Tao, MU Ping, Zi-chao LI, and Yang Ling. 2013. "Quantitative trait loci for mercury tolerance in rice seedlings". *Rice Sci.* 20, No. 3: 238-242.

Weber, Michael, Aleksandra Trampczynska, and Stephan Clemens. 2006. "Comparative transcriptome analysis of toxic metal responses in *Arabidopsis thaliana* and the Cd^{2+}- hypertolerant facultative metallophyte Arabidopsis halleri". *Plant, Cell & Environment* 29, No. 5: 950-963.

Whiting, Steven N, Jonathan R Leake, Steve P McGrath, and Alan JM Baker. 2000. "Positive responses to zinc and cadmium by roots of the hyperaccumulator *Thlaspi caerulescens*". *New Phytologist* 145, No. 2: 199-210.

Wilkinson, Sally, Guzel R Kudoyarova, Dmitry S Veselov, Tatyana N Arkhipova, and William J Davies. 2012. "Plant hormone interactions:

innovative targets for crop breeding and management". *Journal of Experimental Bo*tany 63, No. 9: 3499-3509.

Williams, Lorraine E, Jon K Pittman, and JL Hall. 2000. "Emerging mechanisms for heavy metal transport in plants". *Biochimica et Biophysica Acta (BBA)-Biomembranes* 1465, No. 1-2: 104-126.

Wójcik, M, and A Tukiendorf. 2011. "Glutathione in adaptation of *Arabidopsis thaliana* to cadmium stress". *Biologia plantarum* 55, No. 1: 125-132.

Xiong, Jie, Guanfu Fu, Longxing Tao, and Cheng Zhu. 2010. "Roles of nitric oxide in alleviating heavy metal toxicity in plants". *J. of Biochemistry and Biophysics* 497, No. 1-2: 13-20.

Xu, Jin, Wenying Wang, Jianhang Sun, Yuan Zhang, Qing Ge, Liguo Du, Hengxia Yin, and Xiaojing Liu. 2011. "Involvement of auxin and nitric oxide in plant Cd-stress responses". *Plant and soil* 346, No. 1-2: 107-119.

Yadav, SK. 2010. "Heavy metals toxicity in plants: an overview on the role of glutathione and phytochelatins in heavy metal stress tolerance of plants". *South African Journal of Botany* 76, No. 2: 167-179.

Yan, An, Yamin Wang, Swee Ngin Tan, Mohamed Lokman Mohd Yusof, Subhadip Ghosh, and Zhong Chen. 2020. "Phytoremediation: a promising approach for revegetation of heavy metal-polluted land". *Frontiers in Plant Science* 11.

Yang Z, Chu C. 2011. "Towards understanding plant response to heavy metal stress." In Shanker A, Venkateswarlu B (eds.), *Agricultural and Biological Sciences: Abiotic Stress in Plants-Mechanisms and Adaptations*, 59-78.

Yang, Hai-Ling, Ying-Ru Zhao, Cai-Ling Wang, Zhi-Ling Yang, Qing-Yin Zeng, and Hai Lu. 2009. "Molecular characterization of a dehydroascorbate reductase from Pinus bungeana". *Journal of Integrative Plant Biology* 51, no. 11: 993-1001.

Yang, Tianbao, Gul Shad Ali, Lihua Yang, Liqun Du, ASN Reddy, and BW Poovaiah. 2010. "Calcium/calmodulin-regulated receptor-like kinase CRLK1 interacts with MEKK1 in plants". *Plant Signaling & Behavior* 5, No. 8: 991-994.

Yang, Tianbao, Shubho Chaudhuri, Lihua Yang, Liqun Du, and BW Poovaiah. 2010a. "A calcium/calmodulin-regulated member of the receptor-like kinase family confers cold tolerance in plants". *Journal of Biological Chemistry* 285, No. 10: 7119-7126.

Yang, Xiaoe, Ying Feng, Zhenli He, and Peter J Stoffella. 2005. "Molecular mechanisms of heavy metal hyperaccumulation and phytoremediation". *Journal of Trace Elements in Medicine and Biology* 18, No. 4: 339-353.

Yannarelli, Gustavo G, Ana J Fernández-Alvarez, Diego M Santa-Cruz, and María L Tomaro. 2007. "Glutathione reductase activity and isoforms in leaves and roots of wheat plants subjected to cadmium stress". *Phytochemistry* 68, No. 4: 505-512.

Yuan, Hong-Mei, Heng-Hao Xu, Wen-Cheng Liu, and Ying-Tang Lu. 2013. "Copper regulates primary root elongation through PIN1-mediated auxin redistribution". *Plant and Cell Physiology* 54, No. 5: 766-778.

Zhang, Yuexuan, Zhiling Yu, Xinrong Fu, and Chun Liang. 2002. "Noc3p, a bHLH protein, plays an integral role in the initiation of DNA replication in budding yeast". *Cell* 109, No. 7: 849-860.

Zhao, Feng-Yun, Wen Liu, and Shi-Yong Zhang. 2009. "Different Responses of Plant Growth and Antioxidant System to the Combination of Cadmium and Heat Stress in Transgenic and Non-transgenic Rice". *Journal of Integrative Plant Biology* 51, No. 10: 942-950.

Zobel, Richard W, Thomas B Kinraide, and Virupax C Baligar. 2007. "Fine root diameters can change in response to changes in nutrient concentrations". *Plant and Soil* 297, No. 1-2: 243-254.

CHAPTER FIVE

FLOODING AND SUBMERGENCE RESPONSES IN PLANTS

S. DUHAN

Flooding describes the condition of normal dry land being covered with a large amount of water. It has been identified as a growing threat to plants as the frequency and severity of flooding is increasing day by day. This is creating stress in plants either due to waterlogging or submergence, depending on the depth of the water table. Waterlogging sees the presence of too much water around the root zone of the plant. Submergence refers to the complete immersion of the plant in water, or with only a small part of the shoot apex above its surface: complete submergence is used to describe the first condition, while the second condition is known as partial submergence. The conditions of waterlogging and submergence occur in low-lying areas, which are more prone to flooding. The main consequence of both these stresses is the creation of a hypoxic or anoxic environment in the plant's surroundings, which leads to a lack of available oxygen or its complete absence. As oxygen is required for respiration in plants, its unavailability hampers this process and the plant's metabolism shifts from efficient mitochondrial ATP production (O_2 dependent) to inefficient anaerobic substrate level production of ATP (glycolysis, linked to fermentation). As such, anaerobic respiration leads to carbohydrate starvation and ultimately death if the stress is for prolonged duration. Both these stresses are highly detrimental to the growth of plants and this chapter assesses their responses to submergence and flooding.

Introduction

Flooding is a major abiotic stress affecting plant growth and productivity (Fukao et al., 2019). Flooding is defined as waterlogging or submergence depending on the height of the water table (Pradhan and Mohanty, 2013). The condition in which the root zone of the plant is surrounded by excess water is called waterlogging, whereas the term submergence is used when the complete plant is immersed in water or only the apical part of shoot is exposed to the environment, i.e. complete or partial submergence. On the basis of duration, submergence is also of two types: flash flooding and deepwater.

If the plant remains submerged for 1–2 weeks it is known as a flash flood submergence; when water depths of greater than 100 cm persist for months, it is known as deepwater submergence (Kannan et al., 2017). The stress primarily occurs in low lying areas as these are more prone to flooding (Ahmed et al., 2013). Water is essential for plant life, but an excess of water will also negatively affect plants. The major consequence of flooding and submergence is the restriction of gases, like CO_2 and O_2, with lower diffusion rates in water compared to air (Armstrong and Drew, 2002). As a result, the condition of hypoxia (the unavailability of adequate oxygen) or anoxia (a complete absence of oxygen) is created in the plant's surroundings, hampering the process of respiration. This causes the mode of respiration to shift from aerobic to anerobic, leading to lower production of ATP. This is because anerobic respiration only produces 2 ATP molecules per glucose molecule, rather than the 38 produced through aerobic respiration, and the plant wilts even in the presence of excess water (Voesenek et al., 2006). In addition, mineral toxicity has also been identified as a consequence of flooding stress as substances, like the plant hormone ethylene, and products of the anerobic metabolism of soil microorganisms, like Mn^{2+}, Fe^{2+}, S^{2-}, H_2S, and carboxylic acid, start accumulating (McKee and McKelvin, 1993). Flooding also changes the soil pH and redox potential, which further affects plant growth. This situation is worsened by Al and Mn phytotoxicity, Ca deficiency, and reduced mineralization, which affect root metabolism; there have also been a number of reports showing the adverse effect of increased P, K, Cu, and Fe^{3+} on the availability of N, S, and Zn (Drew, 1997). As is the case for submergence, where the whole plant is surrounded by water, photosynthesis is also affected due to a lack of external CO_2 (Pradhan and Mohanty, 2013). As such, no new carbon accumulation can take place in the plant and existing carbohydrates are consumed through anaerobic respiration, which may lead to carbohydrate starvation and ultimately death. Exposure to hypoxic or anoxic conditions also causes the production of ROS, such as superoxide radicals, hydroxyl radicals, and hydrogen peroxide (Duhan et al., 2017a; Duhan et al., 2017b; Upadhyay, 2018). These ROS are very reactive and cause the peroxidation of lipids, proteins, and nucleic acids in the cell, which further causes membrane disintegration (Sharma et al., 2012; Nita and Grzybowski, 2016). Being sessile, plants have to face these extreme conditions and develop morphological, anatomical, and metabolic strategies to cope. This chapter discusses the physiological, morphological, anatomical and metabolic responses of plants to flooding and submergence stress.

Responses of plants to flooding/submergence stress

Flooding and submergence have a range of effects on plants, including physiological, morphological, anatomical, and metabolic changes [**21**].

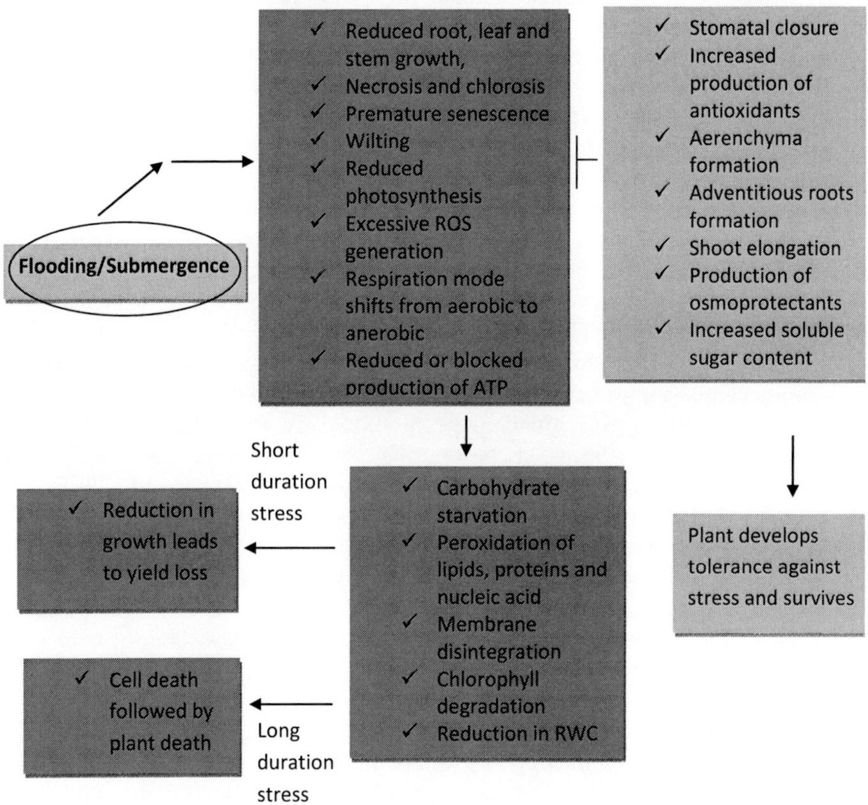

21. Responses of plants to flooding and submergence stress.

Physiological responses

Physiological changes in response to flooding include reduced root, leaf, and stem growth; necrosis and chlorosis resulting in premature senescence; wilting of leaves; stomatal closure; reduced photosynthesis and respiration; and increased susceptibility to disease. Under flooding stress, plants are unable to meet their transpirational demands as the roots are unable to take up water due to disturbed water potential gradients, which leads to wilting. Another reason for wilting is the loss of respiration and ATP synthesis in roots. Stomata close and thus the plant is unable to meet its transpirational demands. Yordanova et al. (2005) found that exposure of barley plants (*Hordeum vulgare* L.) to soil flooding for 2 to 24 hours reduced the net photosynthetic rate, transpiration rate, and stomatal conductance. Stomatal closure started within 2-6 hours of flooding conditions and stomata remained closed up to 24 hours of treatment. An investigation was carried out to determine whether stomatal closure occurs due to decreased leaf water potentials, decreased photosynthetic capacity, and attendant increases in internal CO_2, or from losses of root function, such as cytokinin and gibberellin export, in flooded tomato plants (*Solanum lycopersicum*). The results suggest that stomata closure in flooded plants did not take place in response to short-lived decreases in water potential or to increased CO_2 resulting from impaired PSII photochemistry. Rather, stomatal closure was found to depress CO_2. Stomatal opening was found to be promoted by the presence of well-aerated adventitious roots, suggesting that stomatal closure in the flooded tomato plants was due to a loss of function of root signaling during flooding (Else et al., 2009). Reduced stomatal conductance, transpiration, CO_2 uptake, and leaf elongation in *Ricinus communis* was observed within 2-6 hours of flooding. Continued stomatal closure, slow leaf expansion, and visible wilting was observed in flooded *Ricinus communis* plants as root hydraulic conductance was depressed to 47 % below control values within 2 hours of soil flooding before declining further over the next 21 hours. The smaller root hydraulic conductance temporarily decreased leaf water potentials up to −0.4 MPa at 80 % relative humidity (Else et al., 2001). These studies all suggest that a reduction in water absorption under flooding is a consequence of the reduction in root hydraulic conductivity. It was also found to be associated with acidification of the cell cytoplasm through the gating of aquaporins with an excess of protons, determining the conformational changes of the aquaporins that trigger their closure (Tournarie-Roux et al., 2003; Verdoucq et al., 2008). Several studies have suggested that decreased photosynthesis is a common response to flooding

(Malik et al., 2001; Striker et al., 2005; Mollard et al., 2010). Yordanova and Popova (2001) reported a decrease in the photosynthetic capacity of mesophyll cells in response to continued flooding. This led to a further reduction in photosynthesis, which may have been due to lower leaf chlorophyll content (Yordanova and Popova, 2001; Manzur et al., 2009), reduced activity of carboxylation enzymes, and oxidative damage in the PSII by reactive oxygen species (Yordanova et al., 2004). Similarly, Liao and Lin (1994) found a lower activation level of RUBISCO (the enzyme that catalyzes the initial reaction during CO_2 assimilation) as flooding time increased, reaching 59 % of control values after a week of treatment in *Momordica charantia* (bitter melon). In the same experiment, these authors also observed a reduction in leaf soluble proteins, including RUBISCO (Liao and Lin, 1994). The content of the RUBISCO protein, as well as its activation, are significantly reduced by flooding (Liao and Lin, 2001). In addition, early leaf senescence (Grassini et al., 2007) and a reduction in leaf area may also lead to a drop in carbon fixation (Striker et al., 2005). Contrasting results have been observed in some plants, i.e. the grass *Paspalum dilatatum* and the legume *Lotus tenuis* (Insausti et al., 2001; Striker et al., 2006, 2007b, 2008; Mollard et al., 2008, 2010), although the impact of flooding on plant water differs between species. No major effect of flooding on leaf water potential, stomatal conductance, and transpiration rate has been observed in *Paspalum dilatatum*. No major effect of flooding on leaf water potential, stomatal conductance, and transpiration rate was observed in *Lotus tenuis* for up to a week of flooding treatment, but a visible reduction in leaf water potential, stomatal conductance, and transpiration rate after 15 days of flooding treatment, increasing over time, with reductions of 40 %, 55 %, and 60 % in comparison to control plants. Importantly, when flooding was discontinued, pre-flooded plants of *Lotus tenuis* recovered their previous water status during the first five days and showed stomatal behaviour and transpiration rates similar to control plants until the end of the recovery phase. As such, plant performance during flooding cannot be conclusive in assessments of tolerance—post-flooding recovery also needs to be appraised (Malik et al., 2002; Striker, 2008; Striker et al., 2011b). Decreases in stomatal conductance and transpiration rates have been proposed as mechanisms that regulate the plant water balance and prevent leaf dehydration (Ashraf, 2003; Striker et al., 2005).

Morphological and anatomical responses

Various types of morphological and anatomical changes take place in plants in response to flooding or submergence. Morphological changes include the formation of adventitious roots, which have high porosity and help plants to continue with water and nutrient uptake under flooding conditions, replacing, in some ways, the functions of the older root system (Kozlowski and Pallardy, 1984). It is frequently the case that these adventitious roots are positioned near the soil surface, which has better aeration. The formation of adventitious roots at stem nodes is considered an important morphological adaptation to flooding stress in rice (Upadhyay, 2016). Synthesis of ethylene under flooding conditions, which is thought to increase the sensitivity of plant tissues to auxin, has been suggested as the reason for the formation of adventitious roots (Bertell et al., 1990; Liu and Reid, 1992). Exogenous application of ethylene was found to stimulate the production of adventitious roots without changing the root levels of indole-3-acetic acid (IAA, an auxin) in *Rumex palustris*. This suggests that adventitious rooting is due to increased sensitivity of tissues to auxin and not due to an increase in its concentration (Visser et al., 1995, 1996). A study showed decreased adventitious root formation in wild-type *Nicotiana tabacum* plants on the application of naphthaleneacetic acid (NPA), which is an auxin transport inhibitor (Mc-Donald and Visser, 2003). Morphological changes in shoots included an increase in the insertion angles and length of their aerial organs and an increase in petiole length (Cox et al., 2003, Heydarian et al., 2010). Cox et al. (2003, 2004) found rapid elongation in the shoots of dicotyledonous *Rumex palustris* under flooding. Elongation of submerged tissue may be an important morphological adaptation to tolerate flooding stress in rice (Upadhyay, 2016, 2018). Heydarian et al. (2010) found that submergence induced increased petiole angle and lengthening in *Rumex palustris.* The plant hormone ethylene is thought to be involved in regulating these responses (Vriezen et al., 2000; Heydarian et al., 2010). In graminaceous species, the morphological responses are analogous to those developed by dicots. Insausti et al. (2001) found an increase in the tiller insertion angle as the first morphological response to flooding in the grass *Paspalum dilatatu*, followed by the elongation of the leaf sheaths and the leaf blades. Stem hypertrophy, which is a process that sees the formation of a white spongy tissue homologous to cork outside the phellogen with large volumes of intercellular gas spaces in the stem, has also been identified as a response to flooding (Armstrong et al., 1994). These tissues construct secondary aerenchyma (Shimamura et al., 2010; Teakle et al., 2011) and have been found helpful in increasing air spaces, allowing increased movement of gases between water and plant tissues (Teakle et al., 2011; Upadhyay 2016). Stem hypertrophy has been observed in several plants, such as *Lythrum salicaria* (Stevens et al., 1997), *Lotus uliginosus* (James and Sprent, 1999), *Lotus tenuis* (Striker et al., 2005), *Glycine max* (Shimamura et al., 2010), and *Melilotus siculus* (Teakle et al., 2011). An important morphological trait developed by tolerant woody species is the hypertrophy of lenticels at the stem base (Kozlowski, 1997). It is supposed that these special structures allow the entry of oxygen into shallow roots through aerenchyma and intercellular spaces (Shimamura et al., 2010). Shimamura et al. (2003) performed an experiment and found that blocking the lenticels of waterlogged plants with lanolin resulted in a visible reduction in root aeration; they suggested that lenticels are points of entry for air into the root system. Folzer et al. (2006) found lenticels helpful in maintaining plant water status during flooding by partially supplying water for the shoots and thus replacing less functional roots, which are more likely to develop below water (Parelle et al., 2006). Parent et al. (2008) observed a match in time between the recovery of stomatal conductance and the appearance of hypertrophied lenticels, supporting the idea that they contribute to plant water homeostasis under flooding conditions. Anatomical adaptations include the formation of aerenchyma in the root cortex (Grimoldi et al., 2005; Seago et al., 2005; Striker et al., 2007a) [**22**]. Aerenchyma are continuous gas-filled channels, or much enlarged gas spaces, providing a continuous system of interconnected lacunae of lower resistance for oxygen transport from aerial shoots to submerged roots, thus allowing root growth under anaerobic conditions (Armstrong, 1979; Colmer and Greenway, 2005). Two processes of aerenchyma formation have been recognized: lysigeny and schizogeny (Seago et al., 2005). Lysigeny, which involves the collapse and death of cells in the cortex zone, is the most common process of aerenchyma formation. Schizogeny is a process of aerenchyma generation involving the separation of cells to open up a space (Seago et al., 2005). The condition of hypoxia and ethylene accumulation have both been considered to be behind aerenchyma formation (Jackson et al., 1985; Evans, 2003). Briones et al. (2002) reported that aerenchyma formation allowed oxygen diffusion during anoxia in some rice cultivars. Duhan and Sheokand (2016) observed aerenchyma formation in 20-day-old pigeon pea plant roots 1 and 8 days after removal from an 8-day waterlogging treatment and 1 day after removal from a 12-day waterlogging treatment. Tolerant genotypes (ICPH 2431 and PARAS) showed more aerenchymas compared to sensitive ones (UPAS-120 and HO933). Similar results were observed in

40-day-old pigeon pea plants 1 day after removal from the 8-day waterlogging treatment (Duhan et al., 2017c). Shimamura et al. (2003) reported formation of secondary aerenchyma in the hypocotyl, tap roots, adventitious roots, and root nodules of soybean (*Glycine max*) after three weeks of flooding. No aerenchyma formation was reported under irrigation conditions. In the same experiment, when petroleum jelly was applied to the hypocotyl of plants during a flooded treatment protocol to prevent the entry of atmospheric oxygen into secondary aerenchyma, plant growth, especially in the roots, was sharply inhibited. This suggests that formation of aerenchyma can be considered an adaptive response to flooding stress.

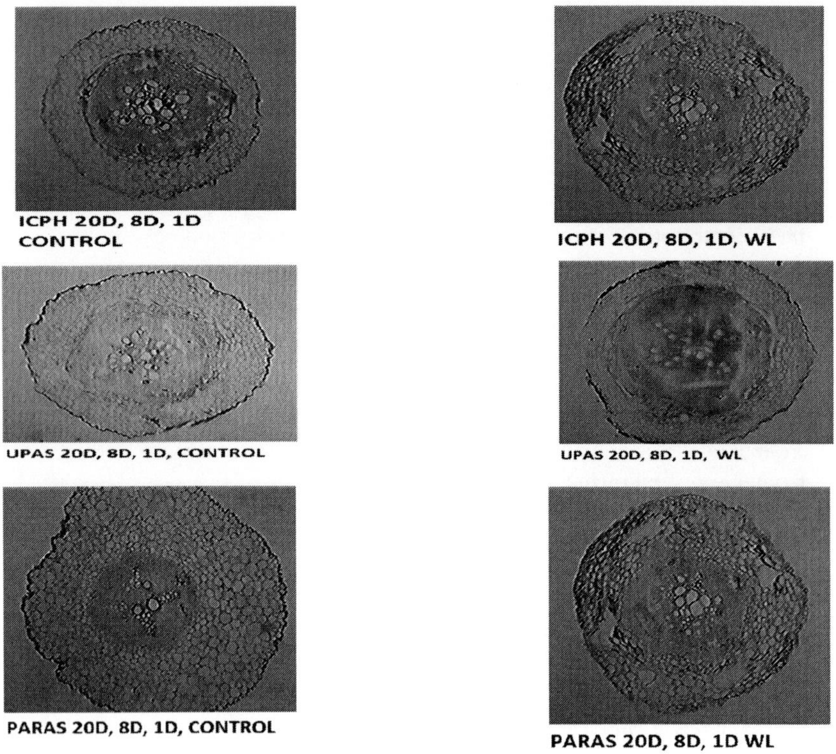

22. Aerenchyma formation in 3 genotypes of 20-day old pigeon pea plants (ICPH 2431, UPAS 120, and PARAS) 1 day after removal from an 8-day waterlogging treatment.

Metabolic responses

The main consequence of flooding or submergence is the creation of a hypoxic or anoxic environment in the area surrounding the plant roots, which limits root respiration. As a resuly, respiration shifts from an aerobic to an anerobic mode. Due to the absence of oxygen, plants are unable to generate ATP through aerobic respiration and to survive have to depend upon the anerobic pathway, which uses reducing sugars (Atwell et al., 2015; Kuai et al., 2016; Duhan et al., 2017d). As photosynthesis is blocked under flooding stress, no carbon accumulation takes place and carbohydrate reserves are used up in anaerobic respiration leading to carbohydrate deficiency, which is thought to be the reason for flooding induced injury (Akhtar and Nazir, 2013). Under hypoxic conditions, a higher amount of sugar content (total, reducing, and non-reducing sugars) was observed in roots of comparatively tolerant genotypes than in those of susceptible genotypes of pigeon pea plants (Hossain and Uddin, 2011; Upadhyay 2016). Ferner et al. (2012) studied the effect of flooding on the carbon metabolism of flooding-tolerant pedunculate oak (*Quercus robur* L.) and flooding-sensitive European beech (*Fagus sylvatica* L.) seedlings. They found decreased soluble carbohydrate concentrations in roots of *Fagus sylvatica* during flooding; however no change in soluble carbohydrate concentrations was observed in *Quercus robur*. At the same time, a decrease in root alcohol dehydrogenase activity was observed in *Fagus sylvatica*, which may have been due to the substrate limitation of alcoholic fermentation in beech roots. Metabolic responses under flooding have also been shown to include the formation of ROS (Upadhyay, 2018). Low photon utilization of flooded plants is thought to be a reason for the production of ROS (Yordanova et al., 2004), which are very reactive and provoke damage to lipid membranes and proteins (Noctor and Foyer, 1998). To manage ROS levels and protect cells, plants have antioxidants, including enzymes (i.e. POX, SOD, GR, CAT, APX, DHAR, and MDHAR) and nonezymatic substances, like ascorbate, glutathione, and tocopherols, which have the ability to scavenge ROS and regenerate antioxidants (Asada, 2006; Voesenek and Bailey-Serres, 2015; Duhan et al., 2019). Upregulation of the antioxidant defense system is often correlated with flood tolerance (Dubey, 2014). Blokhina et al. (2003) observed increased activity of SOD in soybean plant under anoxic stress, which was found to show an increasing trend as the duration of the stress increased. Exogenous ascorbic acid was also found to alleviate postanoxic injury in soybean roots. Increased accumulation of H_2O_2 was observed in the roots and leaves of *Hordeum vulgare* and wheat under hypoxic conditions

(Biemelt et al., 2000). In contrast, wheat plants showed decreased activity of MDHAR and GR enzymes under anoxic conditions (Biemelt et al., 2000). A study showed increased expression of flood-tolerant genes involved in the antioxidant defense and fermentation pathway in maize genotypes when subjected to flooding (Thirunavukkarasu et al., 2013). Similarly, an increase in antioxidative enzyme activity was reported in iris (Monk et al., 1987), wheat (Ushimaru et al., 1997), pigeon pea (Kumutha et al., 2009) welsh onion (Yiu et al., 2008), *Poa pratensis* (Puyang et al., 2015), and *Nymphoides peltata* (Wu et al., 2017). Osmoprotectants, such as glycine betaine, proline, and mannitol, have also been found to help in mitigating flooding stress. Plants accumulate different sugars and amino acids to protect themselves from stress. Glycine betaine is generally found in the chloroplast where it protects thylakoid membranes, while mannitol acts as a free radical scavenger and stabilizes cellular and subcellular structures; it plays a significant role in carbon storage during flooding stress (Arora et al., 2012). Chanu and Sarangthem (2015) subjected rice plants to flooding stress and found increased proline content in the submerged plants. No proline accumulation was observed in the non-submerged plants serving as control. A similar report of proline accumulation in response to flooding in black pepper was reported by Ou et al. (2011). Proline acts as an antidehydrating agent and plays an important role in ROS scavenging, protein stabilization, and DNA stabilization (Arbona et al., 2003, 2008). Accumulation of proline and total sugars was observed in rice cutivars after complete submergence for twenty days (Wangkhem and Kananbala, 2014). On the basis of the above studies, we can suggest that proline accumulation under flooding conditions (through *de novo* synthesis) is an adaptive mechanism to maintain osmoregulation (Chen and Kao, 1993). Hsu et al. (2011) observed increased soluble sugars and amino acids in submerged *Arabidopsis* plants and improved fatty acid and starch catabolism has been suggested as a reason for this increase. Sugar accumulation under flooding conditions in the roots of *Inga marginata* has also been reported (Bender et al., 2016). Qin et al. (2013) reported reduced consumption and enhancement in the accumulation of soluble sugars as a primary strategy to tolerate flooding. Sugars are also thought to act as osmolytes, maintaining cell turgidity and membrane stability, and preventing the degradation of proteins (Sharp et al., 2004).

Conclusions and future perspectives

Flooding is a major abiotic source of stress, which limits plant growth and productivity. The risk of flooding are increasing and it is becoming a great danger to the world's economy. The main consequence of flooding is the creation of hypoxia or anoxia in the surroundings of plants, limiting important processes, such as water uptake, photosynthesis, and respiration. This results in an energy crisis in stressed cells, which triggers tolerant genotypes to develop traits associated with flooding tolerance and improved survival, as summarized in [**21**]. On the other hand, plants of a sensitive nature undergo detrimental changes, such as carbohydrate starvation; the peroxidation of lipids, proteins, and nucleic acid; membrane disintegration; chlorophyll degradation; and a reduction in RWC and cell death followed by the death of the whole plant. So far, we only know a little about the underlying reasons for flooding tolerance and many relevant phenomena are still unknown. Exploring these phenomena will provide fundamental knowledge about the nature of crop responses to flooding. The study of regulatory mechanisms and signaling events responsible for triggering responses to hypoxia or anoxia in plants is an important area of future research.

References

Ahmed, F, Rafii, MY, Ismail, MR, Juraimi, AS, Rahim, HA, Asfaliza, R, Latif, MA. 2013. "Waterlogging Tolerance of Crops: Breeding, Mechanism of Tolerance, Molecular Approaches, and Future Prospects". *BioMed Research International.*

Akhtar, I Nazir, N. 2013. "Effect of waterlogging and drought stress in plants". *International Journal of Water Resources and Environmental Science*, 2(2): 34-40.

Arbona, V, Flors, V, Jacas, J, García-Agustín, P, Gómez-Cadenas, A. 2003. "Enzymatic and non-enzymatic antioxidant responses of Carrizo citrange, a salt-sensitive citrus rootstock, to different levels of salinity". *Plant Cell Physiology*, 44: 388-394.

Arbona, V, Hossain, Z, López-Climent, MF, Pérez-Clemente, RM, Gómez-Cadenas, A. 2008. "Antioxidant enzymatic activity is linked to waterlogging stress tolerance in citrus". *Physiologia Plantarum*, 132: 452-466.

Armstrong, W. 1979. "Aeration in higher plants". *Advances in Botanical Research* 7: 225-332.

Armstrong, W, Drew, M. 2002. “Root growth and metabolism under oxygen deficiency”. In *Plant Roots: The Hidden Half*. Edited by EA Waisel, Y Kafkafi, 729-761 (New York: Marcel Dekker).

Armstrong, W, Strange, ME, Cringle, S, Beckett, PM. 1994. “Microelectrode and modelling study of oxygen distribution in roots”. *Annals of Botany*, 74: 287-299.

Arora, NK, Tewari, S, Singh, S, Lal, N, Maheshwari, DK. 2012. “PGPR for protection of plant health under saline conditions”. In *Bacteria in Agrobiology: Stress Management*, 239-258 (Springer, Berlin Heidelberg).

Asada, K. 2006. “Production and scavenging of reactive oxygen species in chloroplasts and their functions”. *Plant Physiology*, 141: 391-396.

Ashraf, M. 2003. “Relationships between leaf gas exchange characteristics and growth of differently adapted populations of Blue panicgrass (Panicum antidotale Retz.) under salinity or waterlogging”. *Plant Science*, 165: 69-75.

Atwell, B, Timothy, HG, Colmer, D. 2015. “Efficient use of energy in anoxia-tolerant plants with focus on germinating rice seedlings”. *New Phytology*, 206(1): 36-56

Barickman, TC, Simpson, CR, Sams, CE. 2019. “Waterlogging Causes Early Modification in the physiological performance, carotenoids, chlorophylls, proline, and soluble sugars of cucumber”. *Plants*, 8: 160. 10.3390/plants8060160

Bender, B, Capellesso, ES, Lottici, ME, Sentkovski, J, Mielniczki-Pereira, AA, Rosa, LMG, Sausen, TL. 2016. “Growth responses and accumulation of soluble sugars in Inga marginata Wild. (Fabaceae) subjected to flooding under contrasting light conditions”. *Brazilian Journal of Biology*.

Bertell, G, Bolander, E, Eliasson, L. 1990. “Factors increasing ethylene production enhance the sensitivity of root growth to auxins”. *Physiologia Plantarum*, 79: 255-258.

Biemelt, S, Keetman, U, Mock, HP, Grimm, B. 2000. “Expression and activity of isoenzymes of superoxide dismutase in wheat roots in response to hypoxia and anoxia”. *Plant Cell Environment*, 23: 135-144.

Blokhina, O, Virolainen, E, Fagerstedt, V. 2003. “Antioxidants, oxidative damage and oxygen deprivation stress: a review”. *Annals of Botany*, 91: 179-194.

Briones, AM, Okabe, S, Umemiya, Y, Ramsing, NB, Reichardt, W, Okuyama, H. 2002. “Influence of different cultivars on population of ammonia-oxidizing bacteria in the root environment of rice”. *Applied Environmental Microbiology*, 68: 3067-3075.

Cao, YY., Duan, H, Yang, LN, Wang, ZQ, Liu, LJ, Yang, JC. 2009. "Effect of high temperature during heading and early filling on grain yield and physiological characteristics in *indica* rice". *Acta Agronomica Sinica*, 35: 512-521.

Chanu, WS, Sarangthem, K. 2015. "Changes in proline accumulation, amino acid, sugar and chlorophyll content in leaf and culm of Phourel-amubi, a rice cultivar of Manipur in response to flash flood". *Indian Journal of Plant Physiology*, 20: 10-13.

Chen, CT, Kao, CH. 1993. "Osmotic stress and water stress have opposite effects on putrescine and proline production in excised rice leaves". *Plant Growth Regulator*, 13: 197-202.

Colmer, TD, Greenway, H. 2005. "Oxygen transport, respiration, and anaerobic carbohydrate catabolism in roots in flooded soils". In *Plant Respiration: From Cell to Ecosystem.* Edited by H Lambers, M Rivas–Carbo, 137-158. (Springer, Netherlands).

Cox, MCH, Benschop, JJ, Vreeburg, RAM, Wagemaker, CAM, Moritz, T, Peeters, AJM, Voesenek, LACJ. 2004. "The roles of ethylene, auxin, abscisic acid, and gibberellin in the hyponastic growth of submerged *Rumex palustris* petioles". *Plant Physiology*, 136: 2948-2960.

Cox, MCH, Millenaar, FF van Berkel, YEM, Peeters, AJM, Voesenek, LACJ. 2003. "Plant Movement. Submergence-induced petiole elongation in *Rumex palustris* depends on hyponastic growth". *Plant Physiology*, 132: 282-291.

Drew, MC. 1997. "Oxygen deficiency and root metablism: injury and acclimation under hypoxia and anoxia". *Annual Review of Plant Physiology and Plant Molecular Biology*, 48: 223-250.

Dubey, NK. 2014. "Plants as a Source of Natural Antioxidants", CABI.

Duhan, S, Kumari, A, Lal, M, Sheokand, S. 2019. "Oxidative stress and antioxidant defense under combined waterlogging and salinity stresses" In *Reactive Oxygen, Nitrogen and Sulphur Species in Plants*. Edited by Mirza Hasanuzzaman, 113-142 (John Wiley and Sons Ltd).

Duhan, S, Sheokand, S. 2016. "Effects of waterlogging, salinity and combination at different stages of development on survival and root anatomy of pigeon pea (*Cajanus cajan* (L.) Millsp.) genotypes". *Indian Journal of Ecology*, 4(2): 564-568.

Duhan, S, Kumari, A, Sheokand, S. 2017a. "Effect of waterlogging and salinity on antioxidative system in pigeon pea plant leaves at different stages of development". *Research on Crops*, 18(3): 559-568.

Duhan, S, Sheokand, S, Kumari, A. 2017b. "Oxidative stress and antioxidative enzymes activity in pigeon pea leaves at different stages of development under waterlogging, salinity and combined stress of waterlogging and salinity". *Journal of Food Legumes*, 30(2): 59-64.

Duhan, S, Sheokand, S, Kumari, A. 2017d. "Waterlogging, salinity and combined stress a major problem in pigeon pea (*Cajanus cajan* L. Millsp.)". *International Journal of Bioresources and Stress Manage*, 8(2): 291-298.

Duhan, S, Sheokand, S, Kumari, A, Sharma, N. 2017c. "Independent and interactive effects of waterlogging and salinity on carbohydrate metabolism and root anatomy in pigeon pea genotypes at different growth stages". *Indian Journal of Agricultural Research*, 51(3): 197-205.

Else, MA, Coupland, D, Dutton, L, Jackson, MB. 2008. "Decreased root hydraulic conductivity reduces leaf water potential, initiates stomatal closure and slows leaf expansion in flooded plants of castor oil (*Ricinus communis*) despite diminished delivery of ABA from the roots to shoots in xylem sap".

Else, MA, Janowiak, F, Atkinson, CJ, Jackson, MB. 2008. "Root signals and stomatal closure in relation to photosynthesis, chlorophyll *a* fluorescence and adventitious rooting of flooded tomato plants". *Annals of Botany*, 103(2): 313-323.

Evans, DE. 2003. Aerenchyma formation. *New Phytologist*, 161, 35-49.

Ferner, E, Rennenberg, H, Kreuzwieser, J. 2012. "Effect of flooding on carbohydrate metabolism of flood-tolerant (*Quercus robur*) and non-tolerant (*Fagus sylvatica*) tree species". *Tree Physiology*, 32(2): 135-145.

Ferner, E, Rennenberg, H, Kreuzwieser, J. 2012. "Effect of flooding on C metabolism of flood-tolerant (*Quercus robur*) and non-tolerant (*Fagus sylvatica*) tree species". *Tree Physiology*, 32: 135-145.

Folzer, H, Dat, JF, Capelli, N, Rieffel, D, Badot, PM. 2006. "Response of sessile oak seedlings to flooding: an integrated study". *Tree Physiology*, 26: 759-766.

Fukao, T, Barrera-Figueroa, BE, Juntawong, P, Julián, Mario Peña-Castro JM. 2019. "Submergence and Waterlogging Stress in Plants: A Review *Highlighting Research Opportunities and Understudied Aspects*". *Frontiers of Plant Science*.

Grassini, P, Indaco, GV, López Pereira, M, Hall, AJ, Trápani, N. 2007. "Responses to short–term waterlogging during grain filling in sunflower". *Field Crops Research*, 101: 352-363.

Grimoldi, AA, Insausti, P, Vasellati, V, Striker, GG. 2005. "Constitutive and plastic root traits and their role in differential tolerance to soil flooding among coexisting species of a lowland grassland". *International Journal of Plant Sciences*, 166: 805-813.

Heydarian, Z, Sasidharan, R, Cox, MCH, Pierik, R, Voesenek, LACJ, Peeters, AJM. 2010. "A kinetic analysis of hyponastic growth and petiole elongation upon ethylene exposure in *Rumex palustris*". *Annals of Botany*, 106: 429-435.

Hossain, MA, Uddin, SN. 2011. "Mechanisms of waterlogging tolerance in wheat, morphological, and metabolic adaptations under hypoxia or anoxia". *Australian Journal of Crop Science*, 5(9): 1094-1101.

Hsu, FC, Chou, MY, Peng, HP, Chou, SJ, Shih, MC. 2011. "Insights into hypoxic systemic responses based on analyses of transcriptional regulation in Arabidopsis". *PLoS One*, 6 (12): 288.

Insausti, P, Grimoldi, AA, Chaneton, EJ, Vasellati, V. 2001. "Flooding induces a suite of adaptive plastic responses in the grass *Paspalum dilatatum*". *New Phytologist*, 152: 291-299.

Jackson, MB, Fenning, TM, Drew, MC, Saker, LR. 1985. "Stimulation of ethylene production and gas-space (aerenchyma) formation in adventitious roots of *Zea mays* L. by small partial pressures of oxygen". *Planta,* 165: 486-492.

James, EK, Sprent, JI. 1999. "Development of N_2–fixing nodules on the wetland legume *Lotus uliginosus* exposed to conditions of flooding". *New Phytologist*, 142: 219-231.

Kannan, E, Paliwal, A, Sparks, A. 2017. "Spatial and Temporal Patterns of Rice Production and Productivity" In *The Future Rice Strategy for India.* Edited by Samarendu Mohanty, P Chengappa, Mruthunjaya Hedge, JK Ladha, Sampriti Baruah, Elumalai Kannan, AV Manjunatha (Academic Press), 39-68.

Kozlowski, TT, & Pallardy, SG. "1984 Effects of flooding on water, carbohydrate and mineral relations". In *Flooding and plant growth.* Edited by TT Kozlowski (Academic Press Inc., Orlando, Florida) 165-193.

Kozlowski, TT. 1997. "Responses of woody plants to flooding and salinity". *Tree Physiology Monograph 1.*

Kuai, J, Chen, Y, Wang, Y, Meng, Y, Chen, B, Zhao, W, Zhou, Z. 2016. "Effect of waterlogging oncarbohydrate metabolism and the quality of fiber in cotton (*Gossypium hirsutum* L.)". *Frontiers in Plant Science*, 7: 877.

Kumutha, D, Ezhilmathi, K, Sairam, RK, Srivastava, GC, Deshmukh, PS, Meena, RC. 2009. "Waterlogging induced oxidative stress and antioxidant activity in pigeon pea genotypes". *Biologia Plantarum*, 53: 75-84.

Liao, CT, Lin, CH. 1994. "Effect of flooding stress on photosynthetic activities of *Momordica charantia*". *Plant Physiology and Biochemistry*, 32: 479-485.

Liao, CT, Lin, CH. 2001. "Physiological adaptation of crop plants to flooding stress". *Proceedings of the National Science Council, Republic of China Part B*, 25: 148-157.

Liu, JH, & Reid, DM. 1992. "Auxin and ethylene-stimulated advantitious rooting in relation to tissue sensitivity to auxin and ethylene production in sunflower hypocotyls". *Journal of Experimental Botany* 43: 1191-1198.

Malik, AI, Colmer, TD, Lambers, H, Schortemeyer, M. 2001. "Changes in physiological and morphological traits of roots and shoots of wheat in response to different depths of waterlogging". *Australian Journal of Plant Physiology*, 28: 1121-1131.

Malik, AI, Colmer, TD, Lambers, H, Setter, TL, Schortemeyer, M. 2002. "Short–term waterlogging has long–term effects on the growth and physiology of wheat". *New Phytologist*, 153: 225-236.

Manzur, ME, Grimoldi, AA, Insausti, P, Striker GG. 2009. "Escape from water or remain quiescent? Lotus tenuis changes its strategy depending on depth of submergence". *Annals of Botany*, 104, 1163-1169.

McDonald, MP, Visser, EJW. 2003. "A study of the interaction between auxin and ethylene in wildtype and transgenic ethylene insensitive tobacco during adventitious root formation induced by stagnant root zone conditions". *Plant Biology*, 5: 550-556.

McKee, WH, McKevlin, MR. 1993. "Geochemical processes and nutrient uptake by plants in hydric soils". *Environmental Toxicology Chemistry*, 12: 2197-2207.

Mollard, FPO, Striker, GG, Ploschuk, EL, Insausti, P. 2010. "Subtle topographical differences along a floodplain promote different plant strategies among *Paspalum dilatatum* subspecies and populations". *Australian Ecology*, 35: 189-196.

Mollard, FPO, Striker, GG, Ploschuk, EL, Vega, AS, Insausti, P. 2008. "Flooding tolerance of *Paspalum dilatatum* (Poaceae: Paniceae) from upland and lowland positions in a natural grassland". *Flora*, 203: 548-556.

Monk, LS, Fagerstedt, KV, Crawford, RM. 1987. "Superoxide dismutase as an anaerobic polypeptide a key factor in recovery from oxygen deprivation in Iris pseudacorus?" *Plant Physiology*, 85: 1016-1020.

Nita, M, Grzybowski, A. 2016. "The Role of the Reactive Oxygen Species and Oxidative Stress in the Pathomechanism of the Age-Related Ocular Diseases and Other Pathologies of the Anterior and Posterior

Eye Segments in Adults". *Oxidative Medicine and Cellular Longevity.* 10.1155/2016/3164734

Noctor, G, Foyer, CH. 1998. "Ascorbate and glutathione, keeping active oxygen under control". *Annual Review Plant Physiology and Plant Molecular Biology*, 49: 249-279.

Ou, LJ, Dai, XZ, Zhang, ZQ, Zou, XX. 2011. "Responses of pepper to waterlogging stress". *Photosynthetic*, 49: 339-345.

Parelle, J, Roudaut, JP, Ducrey, M. 2006. "Light acclimation and photosynthetic response of beech (*Fagus sylvatica* L.) saplings under artificial shading or natural mediterranean conditions". *Annals of Forest Science,* 63: 257-266.

Parent, C, Capelli, N, Dat, J. 2008. "Reactive oxygen species, stress and cell death in plants". *Comptes Rendus–Biologies*, 331: 255-261.

Pradhan, C, Mohanty, M. 2013. "Submergence Stress: Responses and adaptations in crop plants". In *Molecular Stress Physiology of Plants.* Edited by Gyana Ranjan Rout, Anath Bandhu Das. 331-357 (Springer India).

Puyang, X, An, M, Xu, L, Han, L, Zhang, X. 2015. "Antioxidant responses to waterlogging stress and subsequent recovery in two Kentucky bluegrass (*Poa pratensis* L.) cultivars". *Acta Physiologia Plantarum*, 37: 197.

Qin, X, Li, F, Chen, X, Xie, Y. 2013. "Growth responses and non-structural carbohydrates in three wetland macrophyte species following submergence and de-submergence". *Acta Physiologia Plantarum*, 35: 2069-2207.

Seago, JL, Marsh, LC, Stevens, KJ, Soukup, A, Vortubová, O, Enstone, DE. 2005. "A reexamination of the root cortex in wetland flowering plants with respect to aerenchyma". *Annals of Botany*, 96: 565-579.

Sharma, P, Bhushan, A, Rama, J, Dubey, S, Pessarakli, M. 2012. "Reactive Oxygen Species, Oxidative Damage, and Antioxidative Defense Mechanism in Plants under Stressful Conditions".

Sharp, RE, Poroyko, V, Hejlek, LG, Spollen, WG, Springer, GK, Bohnert, HJ, Nguyen, HT. 2004. "Root growth maintenance during water deficits: physiology and functional genomics". *Journal of Experimental Botany*, 55: 2343-2351.

Shimamura, S, Mochizuki, T, Nada, Y, Fukuyama, M. 2003. "Formation and function of secondary aerenchyma in hypocotyl, roots and nodules of soybean (*Glycine max*) under flooded conditions". *Plant and Soil*, 251: 351-359.

Shimamura, S, Mochizuki, T, Nada, Y, & Fukuyama, M. 2003. "Formation and function of secondary aerenchyma in hypocotyl, roots and nodules of soybean (*Glycine max*) under flooded conditions". *Plant and Soil*, 251: 351-359.

Stevens, KJ, Peterson, RL, & Stephenson, GR. 1997. "Morphological and anatomical responses of Lythrum salicaria L. (purple loosestrife) to an imposed water gradient". *International Journal of Plant Sciences* 158: 172-183.

Striker, GG, Insausti, P, Grimoldi, AA. 2007b. "Effects of flooding at early summer on plant water relations of Lotus tenuis". *Lotus Newsletter*, 37: 1-7.

Striker, GG, Insausti, P, Grimoldi, AA, Vega, AS. 2007a. "Trade-off between root porosity and mechanical strength in species with different types of aerenchyma". *Plant, Cell and Environment*, 30: 580-589.

Striker, GG, Insausti, P, Grimoldi, AA. 2008. "Flooding effects on plant recovery from defoliation in the grass *Paspalum dilatatum* and the legume *Lotus tenuis*". *Annals of Botany*, 102: 247-254.

Striker, GG, Insausti, P, Grimoldi, AA, León, RJC. 2006. "Root strength and trampling tolerance in the grass *Paspalum dilatatum* and the dicot *Lotus glaber* in flooded soil". *Functional Ecology*, 20: 4-10.

Striker, GG, Insausti, P, Grimoldi, AA, Ploschuk, EL, Vasellati, V. 2005. "Physiological and anatomical basis of differential tolerance to soil flooding of *Lotus corniculatus* L. and *Lotus glaber* Mill". *Plant and Soil*, 276: 301-311.

Striker, GG, Izaguirre, RF, Manzur, ME, Grimoldi, AA. 2011b. "Different strategies of Lotus japonicus, *Lotus corniculatus* and *Lotus tenuis* to deal with complete submergence at seedling stage". *Plant Biology*. doi:10.1111/j.1438–8677.2011.00493.x

Teakle, NL, Armstrong, J, Barrett–Lennard, EG, Colmer, TD. 2011. "Aerenchymatous phellem in hypocotyl and roots enables O_2 transport in *Melilotus siculus*". *New Phytologist*, 190: 340-350.

Thirunavukkarasu, N, Hossain, F, Mohan, S, Shiriga, K, Mittal, S, Sharma, R, Singh, RK, Gupta, HS. 2013. "Genome-wide expression of transcriptomes and their co-expression pattern in subtropical maize (*Zea mays* L.) under waterlogging stress". *PLoS One* 8, e70433.

Tournaire–Roux, C, Sutka, M, Javot, H, Gout, E, Gerbeau, P, Luu, DT, Richard Bligny, R, Maurel, C. 2003. "Cytosolic pH regulates root water transport during anoxic stress through gating of aquaporins". *Nature*, 425: 393-397.

Upadhyay, RK. 2018. "Oxidative Injury and its Detoxification in Rice Plants after Submergence Stress". *Proceedings of the National Academy of Sciences India Section B-Biological Sciences*. Springer India, 88: 15-21.

Upadhyay RK. 2016. "How rice (Oryza sativa L.), a semi-aquatic plant adapt to natural flood or submerged condition? A physiological perspective". *Sains Malaysiana*, 45(6): 879-882.

Ushimaru, T, Maki, Y, Sano, S, Koshiba, K, Asada, K, Tsuji, H. 1997. "Induction of enzymes involved in the ascorbate-dependent antioxidative system, namely, ascorbate peroxidase, monodehydroascorbate reductase and dehydroascorbate reductase, after exposure to air of rice (*Oryza sativa*) seedlings germinated under water". *Plant Cell Physiology*, 38: 541-549.

Verdoucq, L, Grondin, A, Maurel, C. 2008. "Structure-function analyses of plant aquaporin AtPIP2;1 gating by divalent cations and protons". *The Biochemical Journal*, 415: 409-416.

Visser, EJW, Cohen, JD, Barendse, GWM, Blom, CWPM, Voesenek, LACJ. 1996. "An ethylene–mediated increase in sensitivity to auxin induces adventitious root formation in flooded *Rumex palustris*". *Plant Physiology*, 112: 1687-1692.

Visser, EJW, Heijink, CJ, Van Hout, KJGM, Voesenek, LACJ, Barendse, GWM, Blom, CWPM. 1995. "Regulatory role of auxin in adventitious root formation in two species of *Rumex*, differing in their sensitivity to waterlogging". *Physiologia Plantarum*, 93: 116-122.

Voesenek, LACJ, Colmer, TD, Pierik, R, Millenaar, FF, Peeters, AJM. 2006. "How plants cope with complete submergence". *New Phytologist*, 170: 213-226.

Voesenek, LA, Bailey-Serres, J. 2015. "Flood adaptive traits and processes: an overview". *New Phytology*, 206: 57-73.

Vriezen, WH, De Graaf, B, Mariani, C, Voesenek, LACJ. 2000. "Submergence induces expansin gene expression in flooding-tolerant *Rumex palustris* and not in flooding intolerant *R. acetosa*". *Planta*, 210: 956-963.

Wangkhem, SC, Kananbala, S. 2014. "Changes in proline accumulation, amino acids, sugar and chlorophyll content in leaf and culm of phourel amubi, a rice cultivar of Manipur in response to flash flood". *Indian Journal Plant Physiology*, 20: 10-13.

Wu, J, Zhao, HB, Yu, D, Xu, X. 2017. "Transcriptome profiling of the floating-leaved aquatic plant *Nymphoides peltata* in response to flooding stress". *BMC Genomics*, 18: 119.

Yiu, JC, Liu, CW, Kuo, CT, Tseng, MJ, Lai, YS, Lai, WJ. 2008. "Changes in antioxidant properties and their relationship to paclobutrazol-induced flooding tolerance in Welsh onion". *Journal of Science, Food and Agricultue*, 88: 1222-1230.

Yordanova, R, Christov, K, Popova, L. 2004. "Antioxidative enzymes in barley plants subjected to soil flooding". *Environmental & Experimental Botany*, 51: 93-101.

Yordanova, RY, Uzunova Losanka, AN, Popova, P. 2005. "Effects of short-term soil flooding on stomata behaviour and leaf gas exchange in barley plants". *Biologia Plantarum*, 9(2): 317-319. 10.1007/s10535-005-7319-6

Yordanova, RY, Popova, LP. 2001. "Photosynthetic response of barley plants to soil flooding". *Photosynthetica*, 39: 515-520.

CHAPTER SIX

THE IMPACT OF PLASTICS ON PLANT GROWTH, AGRICULTURAL CROPS, AND POSSIBLE REMEDIES

ND DHANRAJ, M THAMPI AND MS JISHA

Plastics are widely used due to the variety of benefits they offer. Plastics play an important role in our daily lives, as well as the global economy. Each year, at least 350 to 400 million tons of plastic is produced globally. The application of plastic products have increased modern agricultural practices; this has been termed 'plasticulture' and has enhanced the efficient functioning of agriculture. Applications like plastic mulching of agricultural fields have benefits such as reduced weed growth, reduced evaporation of soil water, and earlier and higher yields, etc. It is well-known that plastics stay in the environment for a long time, as they do not decompose easily. As such, apart from their benefits, the accumulation of plastics poses a serious threat to the environment. Plastic residues may adversely affect the soil ecosystem, harm soil organisms and plants, and reduce soil quality. Plants are the primary producers of an ecosystem and play an important role in maintaining it. Micro and macro plastics adversely affect the soil ecosystem and thereby plant growth. It has already been documented that microplastics have disturbed the growth of some plants. Micro/nano plastics affect soil properties and plant traits. Plastics have serious impacts on the key functions of soil ecosystems, such as soil microbial activity and nutrient cycling, and affect plant growth. Root growth in plants and the movement and transformation of water and nutrients in the soil may be disrupted due to plastics in the topsoil. If this happens in agricultural fields, crop yields will be reduced. This review primarily focuses on efforts to address the problems caused by the extensive use of plastics in plant growth and agriculture; how plastics affect plant growth and crops yield; how they affect the soil ecosystem, which affects plant life; and various aspects of the remediation of plastic pollution to reduce its adverse effects on plant life.

Introduction

Plastics were first used in agriculture in 1948: Prof E. M. Emmert had no money to buy a glasshouse, so he covered a wooden structure with cellulose acetate film. Later, he replaced this with polyethylene film. Over the last few decades, it has been found that plastics adversely affect aquatic ecosystems, and this is true in the case of terrestrial ecosystems (Rillig, 2012).

In the case of the aquatic phase, the effect of plastic is difficult to see with the naked eye when compared to soil. In terrestrial ecosystems, microplastics are already found in agricultural fields, cities, and industrial areas, as well as remote areas (Zhang and Liu, 2018, Piehl et al., 2018; Fuller and Gautam, 2016; Scheurer and Bigalke, 2018). The decomposition of microplastics in soil is not yet fully understand; the assumption is that this material is persistent and will thus accumulate (Rillig, 2012) [**23**].

23. Hypothesized effects of different types of microplastic particles on plants and potential food safety concerns for crops (Rillig et al., 2019).

Microplastic type	**Major hypothesized effect pathway**	**Expected effect size and direction (+, −) for plant growth**	**Potential concern for food safety (in an agricultural context)**
Beads, fragments	Effects similar to minor changes in soil texture	Minor	Minimal
Fibers	Soil structure, bulk density changes	Large (+)	Unclear
Films	Increased soil water evaporation	Intermediate (−)	Unclear
Biodegradable	Nutrient immobilization in soil (short-term)	Intermediate (−)	Decreased nutrient contents

Nanoplastic	Toxicity to plant roots and soil microbiota	Minimal to intermediate (−)	Plastic ingestion (uptake), especially for belowground plant portions

Although limited research has been carried out in this area, the results to date are concerning: fragments of plastic have been found practically all over the world and can trigger many kinds of adverse effects. Studies have shown that one third of all plastic waste ends up in soil or freshwater. Most of this plastic disintegrates into particles smaller than 5 millimetres, known as microplastics, and these break down further into nanoparticles (less than 0.1 micrometre in size). Many of these particles enter the food chain. In this chapter, we discuss the application of plastics in everyday life and agriculture, and the ways in which microplastics can impact plant growth, both positively and negatively. These effects depend on the plant species and type of plastic, and are likely to translate to changes in plant community composition and perhaps primary production. Our framework will serve to guide ongoing and future research on this important topic.

Application of plastics in daily life and agriculture: an overview

Even though today there are limitations on plastic use, the use of plastics has so penetrated our lives that it cannot be stopped. The diversity of polymers and variety of material properties they have made them important in medical and technological advances, energy savings, and other areas. All aspects of daily life, including transport, telecommunications, clothing, footwear, and packaging materials, have seen the production of plastics increase substantially over the last 60 years (Thompson et al., 2009). The salient features, such as lightness, strength, workability, and low cost, make plastics attractive to all productive sectors, including agriculture.

Although hundreds of plastic materials are available commercially, only a few of them qualify as thermoplastic commodities with high volume and low price. This group includes polypropylene and polyethylene [**24**]. These cost-effective polymers are used in flexible barriers, film pouches, stackable crates for transport and storage, caps and closures for containers, and thin-walled containers for the food industry etc. (Andrady and Neal,

2009). Meeker et al. (2009) report that plastics are mixed with additives to enhance performance, including inorganic fillers, thermal stabilizers, plasticizers, fire retardants, and UV stabilizers. Some of these additives are chemically toxic and will produce breakdown products such as phthalates and bisphenol A (BPA), which have adverse effects on human and animal populations.

24. Polymer materials used in everyday life.

POLYMERS	ABBREVIATIONS	COMMON APPLICATIONS
Polyethylene	PE	Packaging, plastic bags and bottles
Polyvinylchloride	PVC	Pipes, windows, and door frames
Polystyrene	PS	License plate frames, CDs etc.
Polyacrylamide	PA	Water purifiers, paper coatings, cosmetic additives, and contact lenses
Polycarbonate	PC	Bullet proof glass, headlamp lenses

Agricultural applications of plastics

The widespread use of plastic in agriculture began in the USA where it was used to cover little greenhouses with cellophane. Picuno (1994) reported that many countries had expanded the use of plastics in agriculture by replacing traditional materials, such as glass to cover greenhouses and paper or straw for soil mulching. The most important applications in agriculture are for greenhouses, mulching, polytunnels, plastic reservoirs, irrigation systems such as filling and spray cones, tape, and the collection, handlings and transport of crops. The extensive use of plastics in agriculture is described by Sarcasascia-Mugnoza et al. (2011) [**25**].

25. Overview of plastics used in agriculture (Sarcasascia et al., 2011).

Protected cultivations films	**Nets**	**Packaging**
• Greenhouses and polytunnels • Low tunnels • Mulching • Nusery films • Direct covering • Covering vineyards and orchards	• Anti-hail • Anti-bird • Windbreaks • Shade • Nets for olives and nut picking	• Fertilizer sacks • Agrochemical cans • Containers • Tanks for liquid storage • Crates Others • Silage films • Fumigation Films • Baler twine • Bale wraps • Nursery pots • String and rope

Some reviews have reported the beneficial effects of plastics for agriculture including an increase in yields, earlier harvests, the reduction of herbicide and pesticide consumption, and water conservation. The efficient use of plastics in farmland has been reported to give higher quality crops and a healthier environment. For example, in northern China, to guard crops from cold weather in early spring and late autumn, plastic tunnels are used (Zhang, 2001). Scarascia et al. (2011) found that plastic-based agriculture provides effective solutions for the growing of crops in arid regions through the use of plastic irrigation tapes, piping, and draining systems, increasing the market for plastic films. These plastic films have been used to convert unproductive land into modern, agriculturally productive land, encouraging economic growth in traditionally poor regions. In European countries, the International Committee on Plastic in agriculture (CIPA) has highlighted how these plastic film-based agricultural systems are mainly used for the production of ornamental flowers, and fruit and vegetables, such as tomato, cucumbers, watermelon, and peppers, etc. Sintim and Flury (2017) found that the use of agricultural film in 2012 amounted to 4.4 million tonnes and is expected to grow to 7.4-7.5 million tons by 2020. The United States Department of Agriculture has recognized the use of plastics in crop production and cultivation, such as plastic film mulches, row covers, tunnels, and greenhouses. In general, plastics have the key benefit of saving other agricultural inputs, such as agrochemicals and fertilizers, and increased water savings. Espi et al. (2006) points out that the important agricultural applications of plastics are mulching, greenhouses, walk in tunnels (high tunnels), and low tunnel covers. Each of these applications is described below.

Mulching

Kader et al. (2017) defines mulches as materials applied to the soil surface, covering it. Mulching is a water conservation method that increases water infiltration into the soil, retards soil erosion, and reduces surface runoff. Mulching helps in manipulating the crop-growing environment to increase crop yield and improve product quality. The application of plastic mulch films helps reduce weeds and pests and protect seedlings and shoots through insulation and the prevention of evaporation, thereby maintaining soil temperature and humidity (Sica et al., 2011). However, low density and linear density polyethylene plastic mulch films are not readily biodegradable. Furthermore, they are often contaminated with soil and are not accepted by recycling facilities. Bowersox and Ward (1970) and Banko and Stefani (1991) investigated how different mulch types influence the water movement. For example, black plastic mulches influence the movement of water between soil layers, resulting in soil erosion and affecting plant growth.

Greenhouses and high tunnels

Greenhouses and high/walk-in tunnels are the quantitatively most important applications of plastic films. These are not permanent, but simple constructions that are vented manually by rolling up the sides. The basic versions have no electrical services, or automated heating and ventilation system. They are used to extend the growing season of crops (Espi et al., 2006). Single or double layer films allow a heating energy saving of 25-30 %. Single layer film is used to increase solar transmittance, whereas double layer film helps reduce water condensation, which is harmful for plants, and avoid thermal inversion of the internal air temperature of 2-3 °C (Scarascia et al., 2011). The covering material determines the significant microclimate inside the tunnel or greenhouse by reducing conductive, convective, and radiant losses of energy. Kittas (1999) found that the quality of the covering was important for evaluating its influence on plant growth and development.

Low tunnels

These are structures built just high enough to cover the plant canopy. Thinner films have shorter lifespans and last for less than one agricultural cycle. EVA or EBA polymers are used because they have particular characteristics of transparency, clarity, and thermal insulation. It has been reported that China saw a 15 % annual growth rate over the past decade through the use of plastic films with a thickness of about 60-100 μm that can be easily opened or removed.

Others

Silage

Bisaglia and Tabacco et al. (2011) report that silage production is an important agricultural practice that offers advantages over hay production, such as flexible harvest dates, lower weather dependency, and greater flexibility. After harvest, fruits are stored in a silage pile covered by plastic in order to protect them against rain. This affects the dry matter and quality losses in terms of field application.

Nets

Plastic nets are widely used in fruit farming for protection from hail, wind, snow, or strong rainfall. Shading nets are used for greenhouse applications in various climatic conditions. Castellano and Mugnozza et al. (2008) reported that nets can be used to moderately modify the microenvironment around crop plants. Moreover, they can protect plants from virus carrying insects and birds, as well as help in harvesting and post-harvesting practices. High-density polyethylenes (HDPE) are most widely used as agricultural nets. Scarascia-Mugnozza and Sica et al. (2011) found that, in Italy, 5,300 ton HDPE per year is used for agricultural net production.

Other plastic products used in the agricultural sector

Plastics are also used for irrigation and drainage pipes or tapes; pots for ornamental plants and flowers; nursery containers; soilless culture substrate; strings, bags and containers for agrochemicals. Plastic films are used in the packaging, transport, storage, and sale of agricultural products.

Accumulation of plastics (major source and reasons)

Plastics have transformed our everyday lives and its usage is increasing day by day. It brings new challenges to the environment and human health. Plastic accumulation harms humans, animals, and plants through toxic pollutants. The usage and disposal of plastics are a major concern due to the accumulation of waste in landfills, triggering changes in soil chemistry and affecting soil-plant interaction, agricultural system, and natural habitats, as well as causing physical problems for wildlife. Results can include plastic entanglement, leaching, and the production of toxic material. Barnes et al. (2009) reported that around 10 % of municipal waste is plastic, contaminating a wide range of natural terrestrial, fresh water, and marine habitats.

Plastic production and development accelerated after World War II. Plastics made from fossil fuels are just over a century old. The production and development of thousands of new plastic products has transformed the modern age and life without plastics would be unrecognizable today. Plastics have revolutionized medicine with life-saving devices; made space travel possible; lightened cars and jets, saving fuel and reducing pollution; and saved lives with helmets, incubators, and equipment for clean drinking water. Everyday uses of plastic account for 40 % of plastic production every year. This includes products such as plastic bags and food wrappers, used for minutes or hours, but which will exist in the environment for hundreds of years. Discarded plastics produce contaminants that can affect natural terrestrial, fresh water, and marine habitats.

The main source of synthetic plastics is crude oil. Petrol, paraffin, lubricating oils, and high petroleum gases are all byproducts produced during the refining of crude oil. These gases are broken down into to monomers. Monomers are single molecule chemical substances. Polymerization sees thousands of monomers linked together, forming compounds called polymers. Combining the element carbon with other elements, such as oxygen, hydrogen, chlorine, or fluorine, is how most polymers are formed. Plastics can be found in natural substances or may be manmade and known as synthetic plastics. Natural plastics occur in such things as animal horns, animal milk, insects, plants, and trees [**26**].

26. Plastics found in natural substances.

Animal horns	Casein (glue)
Animal milk	Formaldehyde (glue)
Insects	Shellac (French polishing)
Plants	Cellulose (table tennis balls) Cellulose acetate (cloth, photographic film, handles) Cellophane (wrapping) Bitumen (roads, flat roofs)
Trees	Latex (rubber) Rosin (resin) paint Amber (semi-precious decoration)

Zubris and Richards (2005) reported the contamination of soils with small plastic fragments from the spreading of sewage sludge and fragments of plastics and glass contaminating compost prepared from municipal solid waste. There is a need for active research into the quantities and effects of plastic debris in natural terrestrial habitats, on agricultural land, and in fresh water systems. Harper and Fowler (1987) first reported plastic in the environment in seabird carcasses collected on the shoreline in the early 1960s.

Sewage is an important factor in the distribution of microplastics. In fact, between 80 and 90 % of plastic particles contained in sewage, from garment fibres for example, persist in sludge. Sewage sludge is often applied to fields as fertilizer, meaning that several thousand tons of microplastics end up in our soils each year. Microplastics can even be found in tap water. Chlorinated plastic can release harmful chemicals into the surrounding soil, which can then seep into groundwater or other surrounding water sources, as well as the ecosystem. This can cause a range of potentially harmful effects on animals that drink the water. Generally speaking, when plastic particles break down, they gain new physical and chemical properties, increasing the risk that they will have toxic effects on organisms. The larger the number of potentially affected species and ecological functions, the more likely it is that toxic effects will occur. Chemical effects are especially problematic at the decomposition stage. Additives, such as phthalates and bisphenol A (widely known as BPA) leach out of plastic particles. These additives are known for their hormonal effects and can disrupt the hormone system of vertebrates and invertebrates alike. In addition, nano-sized particles may cause inflammation, traverse cellular barriers, and even cross highly selective membranes, such as the blood-brain barrier and the placenta. Within the cell, they can trigger changes in gene expression and biochemical reactions. While there is clear evidence that these plastics have adverse effects at environmentally relevant concentrations in laboratory studies, there is a need for further research to establish how they affect plant population levels in the natural environment (Zubris and Richards, 2005).

Impacts of plastics on plants

The effects of plastics on plant life are seen in the degradation of soil quality due to plastic accumulation. Plants are influenced by soil properties and there exists a relationship between particular soils and the vegetation cover of that soil. Proper soil conditions are required for successful plant development and for crop production. Poor soil conditions, including salinity and inadequate nutrient concentrations, have adverse effects on the life of plants (Pessarakli and Szabolcs, 1999). An increase in the accumulation of plastic residues hinders the movement of moisture in soil. This reduced moisture movement leads to secondary salinization of the surface soil layer and the increase in salinity stress reduces plant growth (Lu et al., 2009). A previously mentioned impact is the deterioration of the physical properties of the soil due to increased accumulation of plastic residues. The bulk density of the soil increases due to the accumulation of plastics and this increased bulk density results in a slowdown of moisture movement. Consequently, plant growth is retarded (Xie et al., 2007).

Vegetative growth, dry biomass, and the fruit production of plants are all adversely affected by the accumulation of plastic residues in the soil. Water retention and flow, and soil structure and texture are also greatly affected by the accumulation of plastic residues in the soil. As such, negative changes in the soil structure and texture, especially in soil porosity, will significantly reduce water soil movement and also the transport of soil solutes. A decrease in soil porosity is due to an increase in the density of plastic residues or small pieces in the soil, preventing the flow of water and nutrients. This leads to poor retention of water and nutrients, ultimately hindering the normal growth of plants (Zou et al., 2017). Water infiltration into the soil is disrupted by plastic mulches, which decrease surface permeability. This condition leads to an increase in runoff and erosion, thereby affecting soil fertility and plant growth (Chalker, 2007).

Impacts of plastics on seed germination

The germination of seeds and flowering are both adversely affected by the accumulation of plastic residues. Plastic film residues left on the soil surface may wrap seedlings, preventing them from absorbing moisture and germinating. Even though some seeds may germinate in these adverse conditions, they may be unable to establish themselves correctly. Plastic residues in the soil prevent proper root growth, change the soil moisture level, reduce the nutrient surface area, and worsen the soil nutrient concentrations. These all contribute to a decrease in flowering and thus fruiting too (Hegan et al., 2015).

Effects of plastics on the soil nutrient profile and plant growth

The accumulation of plastic residues in the soil has a detrimental effect on its nutrient profile. When plastic film density increases, the soil pH also increases, showing that plastic residues also influence soil chemical properties. Plastic residues reduce the alkaline hydrolysis of nitrogen and available phosphorus and potassium. The availability of soil NPK is very important for plant growth and if the availability of NPK is reduced plant growth will be affected. Decreased water and fertilizer infiltration due to plastic residues in farm land also reduce crop growth (Hegan et al., 2015).

Soil quality is a very important factor for plant growth. Biological factors also influence soil quality. The main biotic factor that determines soil quality is the composition and diversity of the soil's microbial community, which plays an important role in carbon, nitrogen, phosphorus, and potassium cycling. Efficient functioning of these cycles is positively correlated with soil fertility. The activity of enzymes taking part in different nutrient cycles can be significantly reduced due to the inhibitory effect of residual plastic film contaminants. Thus, over time, plastic contamination in the soil will reduce plant growth due to a changed nutrient cycle and low fertility. Soil-beta-glucosidase (S-β-GC), which takes part in the carbon cycle, shows reduced activity in soil covered with plastic. Furthermore, the level of soil urease (S-UE) has been found to be lowered. This reduced S-β-GC activity and S-UE activity will negatively affect soil fertility and ultimately plant growth. Soil organic matter decomposition and carbon fixation by microorganisms are the two main ways in which soil carbon is formed. A reduction in the carbon fixation gene (*cbbl*) and 2 carbon source hydrolase coding genes (β-*glu* and *chi*-A) in plastic covered areas is evidance of reduced microbial activity and low decomposition of soil organic matter, lowering soil fertility and plant growth (Samborska et al., 2004; Zhou et al., 2005; Qian et al., 2007; Chen et al., 2009; Xie et al., 2010; Qian et al., 2018). Denitrification genes, such as *nir*S and *nos*Z, show abundance in copy numbers in soil covered with plastic films, highlighting the increased activity of denitrifiers. This results in increased quantities of nitrates in the soil (Tao et al., 2018; Qian et al., 2018).

Effects of microplastics on plant growth

Microplastics adversely effect plant growth. Microplastics can be considered a soil contaminant. Plastic films (2, 5, and 10 mm size fragments, added at 0.5 % and 1.0 %) may create channels increasing water movement. This causes increased evaporation of water and leads to the drying of the soil, which will adversely affect plant growth and performance (Wan et al., 2019). Microplastics may change the soil structure, which also has many negative effects on plant growth. Another problem caused by these microplastics is nutrient immobilization, especially when biodegradable microplastics reach the soil. This causes a very large carbon : nitrogen (C : N) ratio. As such, biodegradable plastics adversly effect the plant performance parameters and, for example, leaf area may decrease—this is not just a long term effect, but also one that cannot be ignored (Qi et al., 2018). Toxic substances that have been adsorbed or are already present in microplastics can adversly effect plant roots and their symbionts, harming plant growth. Microplastics affect plant performance through various mechanisms (Rillig, 2019) [**27**].

Different plant species in a community may be affected at different degrees by the addition of microplastics. Microplastics can adversely affect plant diversity and community composition and there are several ways by which plants are affected through changes to soil structure, increased soil water evaporation, and reduced soil microflora (Rillig, 2019).

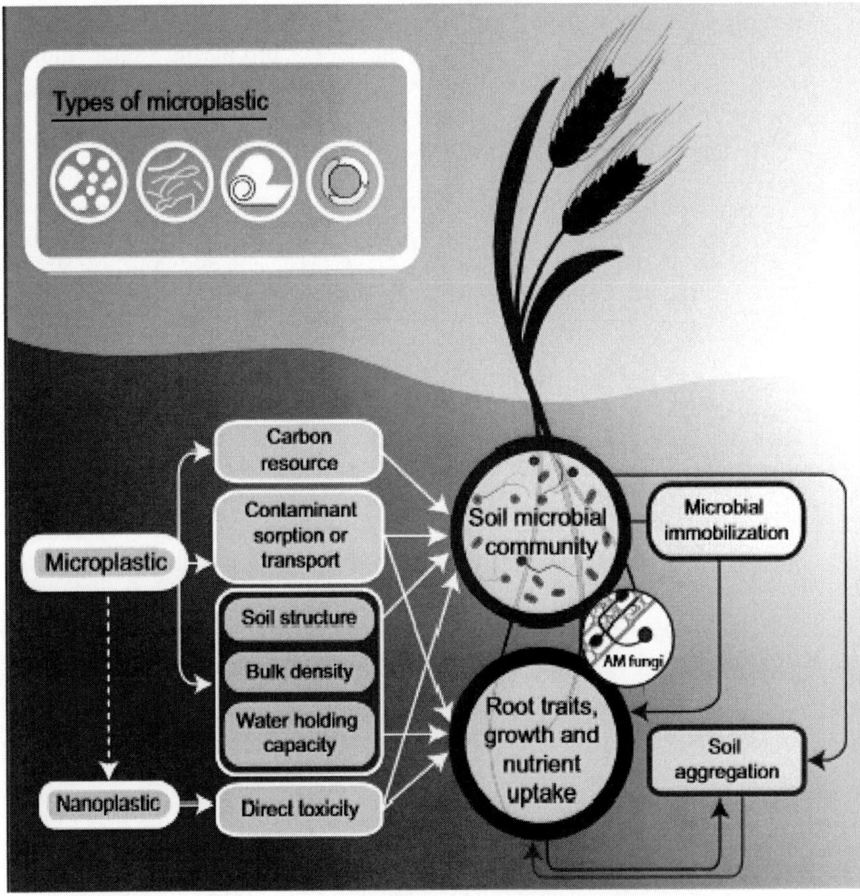

27. Concept diagram showing the various mechanisms via which microplastic particles can affect plant growth (Rillig, 2019).

Remedial measures to mitigate the impacts of plastics on plants and future prospects

Biodegradable plastic mulch films can be used after pretreatment with a culture filtrate of the yeast *Pseudozyma antartica*. This treatment helps to completely degrade the mulch film residue in the soil within 5 weeks, reducing the effects on plants of plastic residues left after mulching (Sameshima et al., 2019). Some policy considerations have to be set for reducing unintended residual plastics in agricultural soil. Compatibility standards currently used to certifying biodegradable plastic mulch film need to take into consideration the effects of plastic mulch film fragments on soil-dwelling microflora and fauna. As such, there is a need to develop new standards (Kershaw, 2015; Brodhagen et al., 2017).

The use of plastic mulch film should be reduced as much as possible; it is a poor choice for mulching as it limits soil water infiltration. There are a wide variety of mulcing materials that, being permeable, do not limit soil water infiltration. We should also use organic mulches because they conserve water more effectively than inorganic mulches (Chalker, 2007).

Moreover, we have to control the flow of undesired plastic waste into forests, agricutural fields, and mangrove swamps, etc. A major land-based source of plastic pollution is mismanaged waste. This can be reduced by improvements in the plastics life-cycle, especially in production, consumption, and disposal through integrated waste management systems (Prata et al., 2019). For the protection of the environment, there is a need not only for sound science, but also for effective policies. In-soil biodegradation standards need to be reworked and should include a reasonable testing range of soil moisture and temperature, crop and soil types, and associated microbial communities; testing should be carried out under selected field conditions, simulating typical farming practices. Furthermore, long term assessments of biodegradable plastics have to be carried out, including assessments of the number, size, character, fate, and transport of mulch film fragments, along with their effects on the soil ecosystem, and the need to encompass repeated mulch laying and tilling over multiple years. However, these kind of long-term studies need to be taken over by federal agencies, industry, academia, and other organizations, supporting research on agriculture and environmental safety (Brodhagen et al., 2017).

Detailed studies need to be undertaken to understand the mechanisms by which conventional and biodegradable pastics affect soil ecosystems and plant growth (Qi et al., 2019). Policy solutions are important and urgently needed, but they also have to be balanced because they should not be so rigid as to restrict innovation in biodegradable plastic films intended for agricultural purposes (Brodhagen et al., 2017).

As mentioned earlier, soil quality is very much related to plant growth. Research into the behavior of microplastics in soil is developing, but there is still a remarkable lack of data. Their characteristics, potential ecological effects, and the underpinning mechanisms of their toxicity remain under-addressed. Evidence about how microplastics affect plant growth is very scarce and a detailed understanding of all those things mentioned are essential for protecting plants. This demands more study into the effects of microplastics on plant growth and their different aspects (Zhu et al., 2019). The effects of plastics on agricultural crops are usually studied, but research should concentrate on other plants too. Also, it is very clear that more research is urgently needed to completely understand the effects of microplastics on the soil plant system and agro-ecosystem (Qi, et al., 2018).

References

Andrady, AL, Neal MA. 2009. "Applications and societal benefits of plastics". *Philosophical Transactions of the Royal Society B: Biological Sciences*, *364* (1526): 1977-1984.

Banko, TJ, MA Stefani. 1991. "Effects of container medium peat content and bed surface on plant growth during capillary irrigation". *J. Environ. Hort*, 9: 33-36.

Bisaglia, C, E Tabacco and G Borreani. 2011. "The use of plastic film instead of netting when tying round bales for wrapped baled silage". *Biosystems Engineering*, 108(1): 1-8.

Bowersox, TW, and WW Ward. 1970. "Black polyethylene mulch - an alternative to mechanical cultivation for establishing hybrid poplars". *Tree Planters' Notes*, 21: 21-24.

Brodhagen, M, Goldberger, JR, Hayes, DG, Inglis, DA, Marsh, TL, and Miles, C. 2017. "Policy considerations for limiting unintended residual plastic in agricultural soils". *Environmental Science & Policy*, 69: 81-84.

Castellano, S, GS Mugnozza, G Russo, D Briassoulis, A Mistriotis, S Hemming and D Waaijenberg. 2008. "Plastic nets in agriculture: a general review of types and applications". *Applied engineering in agriculture*, 24(6): 799-808.

Chalker-Scott, L. 2007. "Impact of mulches on landscape plants and the environment—a review". *Journal of Environmental Horticulture*, 25(4): 239-249.

Chen, H, Yao, J, Wang, F, Choi, MM, Bramanti, E, and Zaray, G. 2009. "Study on the toxic effects of diphenol compounds on soil microbial activity by a combination of methods". *Journal of Hazardous Materials*, 167(1-3): 846-851.

Espi, E, A Salmeron, A Fontecha, Y García and A Real (2006). "Plastic films for agricultural applications." *Journal of Plastic Film & Sheeting*, 22(2): 85-102.

Fuller S, Gautam A. 2016. "A procedure for measuring microplastics using pressurized fluid extraction". *Environmental Science & Technology*, 50: 5774-5780.

Harper, PC, and Fowler, JA. 1987. "Plastic pellets in New Zealand storm-killed prions (Pachyptila spp.)". *Notornis*, 34: 65-70.

Hegan, D, Tong, L, Zhiquan, H, Qinming, S, and Ru, L. 2015. "Determining time limits of continuous film mulching and examining residual effects on cotton yield and soil properties". *Journal of Environmental Biology*, 36(3): 677-684.

Kader, MA, M Senge, MA Mojid and K Ito. 2017. "Recent advances in mulching materials and methods for modifying soil environment." *Soil and Tillage Research*, 168: 155-166.

Kershaw, PJ (ed.) 2015. "Sources, Fate and Effects of Microplastics in the Marine Environment: A Global Assessment". *Rep. Stud. GESAMP*, No. 90.

Kittas C, Baille A, Giaglaras, P. 1999. "Influence of covering material and shading on the spectral distribution of light in greenhouses". *Journal of Agricultural Engineering Research*, 73: 341-351.

Lv, ZY, Niu, LA, Hao, JM, Hu, J, and Sheng, GC. 2009. "Problems faced by ecological environment of agriculture in China and their improving countermeasures". *Chinese Agricultural Science Bulletin*, 25: 218-224.

Meeker, JD, Sathyanarayana, S, and Swan, SH. 2009. "Phthalates and other additives in plastics: human exposure and associated health outcomes". *Phil. Trans. R. Soc. B*, 364: 2097-2113.

Pessarakli, M, and Szabolcs, I. 1999. "Soil salinity and sodicity as particular plant/crop stress factors". In *Handbook of Plant and Crop Stress*, edited by Mohammad Pessarakli, 1-15. New York: Marcel Dekker, Inc.

Picuno, P, Scarascia-Mugnozza, G. 1994. "The management of agricultural plastic film wastes in Italy". In *Proceedings of the International Agricultural Engineering Conference*, edited by VM Salokhe and Gajendra Singh, 797-808. Bangkok (Thailand): Asian Institute of Technology.

Piehl, S, Leibner, A, Loder, MGJ, Dris, R, Bogner, C, Laforsch, C. 2018. "Identification and quantification of macro and microplastics on an agricultural farmland". *Scientific Reports*, 8(1): 1-9.

Prata, JC, Silva, ALP, Da Costa, JP, Mouneyrac, C, Walker, TR, Duarte, AC, and Rocha-Santos, T. 2019. "Solutions and integrated strategies for the control and mitigation of plastic and microplastic pollution". *International Journal of Environmental Research and Public Health*, 16(13): 2411.

Qi, Y, Yang, X, Pelaez, AM, Lwanga, EH, Beriot, N, Gertsen, H, and Geissen, V. 2018. "Macro and micro-plastics in soil-plant system: effects of plastic mulch film residues on wheat (Triticum aestivum) growth". *Science of the Total Environment*, 645: 1048-1056.

Qian, H, Hu, B, Wang, Z, Xu, X, and Hong, T. 2007. "Effects of validamycin on some enzymatic activities in soil". *Environmental Monitoring and Assessment*, 125:1-8.

Qian, H, Zhang, M, Liu, G, Lu, T, Qu, Q, Du, B, and Pan, X. 2018. "Effects of soil residual plastic film on soil microbial community structure and fertility". *Water, Air, & Soil Pollution*, 229(8), 261: 1-11.

Rillig, MC. 2012. "Microplastic in terrestrial ecosystems and the soil?". *Environmental Science & Technology*, 46: 6453–6454.

Rillig, MC, Lehmann, A, de Souza Machado, AA, and Yang, G. (2019). "Microplastic effects on plants". *New Phytologist*, 223(3): 1066-1070.

Samborska, A, Stepniewska, Z, and Stepniewski, W. (2004). "Influence of different oxidation states of chromium (VI, III) on soil urease activity". *Geoderma*, 122(2-4): 317-322.

Sameshima-Yamashita, Y, Ueda, H, Koitabashi, M, and Kitamoto, H. 2019. "Pretreatment with an esterase from the yeast Pseudozyma antarctica accelerates biodegradation of plastic mulch film in soil under laboratory conditions". *Journal of Bioscience and Bioengineering*, 127(1): 93-98.

Scarascia-Mugnozza, G, C Sica and G Russo. 2011. "Plastic materials in European agriculture: actual use and perspectives", 42(3): 15-28.

Scheurer M, Bigalke M. 2018. "Microplastics in Swiss floodplain soils". *Environmental Science & Technology*, 52: 3591-3598.

Sintim, HY, and M Flury. 2017. "Is Biodegradable Plastic Mulch the Solutionto Agriculture's Plastic Problem?" *Environmental Science & Technology*, 51(3): 1068-1069.

Tao, R, Wakelin, SA, Liang, Y, Hu, B, and Chu, G. 2018. "Nitrous oxide emission and denitrifier communities in drip-irrigated calcareous soil as affected by chemical and organic fertilizers". *Science of the Total Environment*, 612: 739-749.

Thompson, RC, Moore, CJ, Vom Saal, FS, and Swan, SH. 2009. "Plastics, the environment and human health: current consensus and future trends". *Philosophical Transactions of the Royal Society B: Biological Sciences*, 364(1526): 2153-2166.

Wan, Y, Wu, C, Xue, Q, and Hui, X. 2019. "Effects of plastic contamination on water evaporation and desiccation cracking in soil". *Science of the Total Environment*, 654:576-582.

Xie, HE, Li, YS, Yang, SQ, Wang, JJ, Wu, XF, and Wu, ZX. 2007. "Influence of residual plastic film on soil structure, crop growth and development in fields". *Journal of Agro-Environment Science*, 26(S1): 153-156.

Xie, HJ, Shi, YJ, Zhang, J, Cui, Y, Teng, SX, Wang, SG, and Zhao, R. 2010. "Degradation of phthalate esters (PAEs) in soil and the effects of PAEs on soil microcosm activity". *Journal of Chemical Technology & Biotechnology*, 85(8): 1108-1116.

Zhang GS, Liu YF. 2018. "The distribution of microplastics in soil aggregate fractions in southwestern China". *Science of the Total Environment*, 642: 12-20.

Zhang ZH, Li JW. 2001. "Present situation and discussion on protected horticulture of China". *China Vegetables*, 3: 1-4.

Zhou, QH, Wu, ZB, Cheng, SP, He, F, and Fu, GP. 2005. "Enzymatic activities in constructed wetlands and di-n-butyl phthalate (DBP) biodegradation". *Soil Biology and Biochemistry*, 37(8): 1454-1459.

Zhu, F, Zhu, C, Wang, C, and Gu, C. 2019. "Occurrence and ecological impacts of microplastics in soil systems: a review". Bulletin of Environmental Contamination and Toxicology, 102(6): 741-749.

Zou, X, Niu, W, Liu, J, Li, Y, Liang, B, Guo, L, and Guan, Y. 2017. "Effects of residual mulch film on the growth and fruit quality of tomato (*Lycopersicon esculentum* Mill.)". *Water, Air, & Soil Pollution*, 228(2): 71.

Zubris, KAV, and Richards, BK. 2005. "Synthetic fibers as an indicator of land application of sludge". *Environ. Pollut.* 138: 201-211.

CHAPTER SEVEN

INTERACTIONAL EFFECTS OF UV-B RADIATION AND DROUGHT STRESS ON PLANTS

B JAMRA, RS BAPNA, S KATARIA AND M JAIN

Due to their immobile nature, plants constantly face conditions of biotic and abiotic stress. These stresses have acute and adverse effects on plant growth and development. Plants are constantly exposed to a broad range of combined abiotic stresses. The majority of experimental work looks at abiotic stress under controlled conditions and the responses of plants to alteration in environmental conditions have generally focused on a single stressor. However, under natural conditions, combinations of two or more abiotic stresses can occur in the field at the same time. These may include conditions such as drought and salinity; drought and chilling; drought and UV; heat and ozone; UV and heavy metals; ozone and pathogens; and high CO_2 and drought. These are common to many agricultural areas around the world and can impact plant development, metabolism, and yield in novel ways that may be different to those caused by different stresses applied individually. To avoid this type of combined abiotic stress, plants have evolved many morphological, physiological, biochemical, and antioxidant mechanisms. Tolerant plants have evolved stomata closure, root growth stimulation, free proline accumulation, and osmolytic deposition in response to multiple stresses. Non-enzymatic and enzymatic antioxidant defense systems have also evolved for protection against drought stress and UV-B stress. Various non-enzymatic compounds, such as glycine betaine, ascorbate, and glutathione (low molecular antioxidants); and antioxidant enzymes, such as superoxide dismutase, catalase, and ascorbate peroxidase; as well as some secondary metabolites, such as polyamins, tochopherol, and cerotenoids, participate in the removal of ROS under increased UV-B and drought stress. The combined effect of drought stress and UV-B radiation induces responses that can be synergistic, additive, or antagonistic in comparison to responses to separate or individual stresses. As such, drought stress may enhance tolerance to UV-B radiation and vice-versa. Hydrogen peroxide, nitric oxide, and some plant hormones participate in the activation of defense mechanisms against both these stress conditions. Here, we will address the combined impact of UV-B and drought stress on plant growth, and morphological, physiological, and biochemical responses and antioxidant defense mechanisms.

Introduction

Owing to their sessile lifestyle, plants are constantly exposed to a broad range of environmental stresses. These can have acute and adverse effects on plant growth and development (Rejeb et al., 2014). Ecological stress factors are divided into two groups: biotic and abiotic. Bacteria, fungi, insects, weeds, viruses, and human activities are all biotic stresses, while high UV radiation, drought, salinity, pesticides, low temperatures (chilling and freezing), O_3, decreased soil oxygen, mineral nutrient deficiency, metal toxicity, and pollutants are all abiotic stresses (Kanojia and Dijkwel, 2018). Among these stressors, abiotic factors are major hazards, particularly in terms of the negative affects on the growth and productivity of crop species. Abiotic stresses are responsible for more than 50 % of plant damage and reduced crop productivity across the world (Mittler et al., 2001). The majority of experimental work on abiotic stress in plants has focused on a single stress response to alteration in controlled environmental conditions (Cushman and Bohnert, 2000; Zhu, 2002). However, under natural conditions, combinations of two or more abiotic stresses can occur simultaneously. These combinations may include conditions such as drought and salinity; drought and chilling; drought and UV; heat and ozone; UV and heavy metals; ozone and pathogens; and high CO_2 and drought. Such combined stress conditions are common to many agricultural areas around the world and can impact plant development, metabolism, and yield in novel ways, which may be different to the results caused by each stress applied independently in the lab. In response to such abiotic stress conditions, plants build up several morphological, physiological, biochemical, metabolic strategies, and defense mechanisms. The majority of stress combinations studied have negative effects on crop yield and biomass production in the field.

Drought stress is the most common environmental constraint due to inadequate rainfall and reduced water availability. It is frequently accompanied by increased UV-B radiation (Ballaré et al., 2011; Bandurska et al., 2013). It has been revealed that the combined effect of high UV-B radiation and drought stress results in the reduction and alteration of plant growth, as well as of morphological, physiological, and biochemical processes (Alexieva et al., 2001). Both these environmental factors act synergistically on the plant's secondary metabolism by increasing the production of flavonoids (Hofmann et al., 2003), which induce the production of cuticular waxes, thus facilitating the reflection of light and better water conservation. Enhanced and ambient UV-B radiation has a significant effect on the photosynthetic efficiency of crop plants due to a reduction in the efficiency of PS II and the activity of RUBISCO, PEP carboxylase, and carbonic anhydrase (Kataria et al., 2014). In contrast, drought stress restricts the productivity of crop plants by affecting photosynthetic processes either directly, or by feedback inhibition at the chloroplast level. The response pattern is regulated by the intensity, duration, and rate of progression of drought stress on plants (Pinheiro and Chaves, 2011). Stomatal characteristics are affected under mild to moderate conditions, reducing biomass, whereas non-stomatal factors can become dominant under severe conditions.

Morphological responses to UV-B and drought

During the co-occurrence of stresses, each stress factor may alleviate or increase the negative effects of the other stress (Mittler, 2006). However, the combined action of drought stress and UV-B stress can transform the response patterns. They can interact in a synergistic or additive manner, but can also interact to significantly increase resistance mechanisms. Plant reactions to combined stress depend on the sensitivity of the species, and the intensity, duration, exposure, and mode of action of the stress factors. In addition to the action of the stresses, sequential and simultaneous protective effects are observed, but the amplification of negative effects may occur simultaneously as well as sequentially with prolonged or severe stress. The responses of plants to abiotic stresses are shown in [**28**].

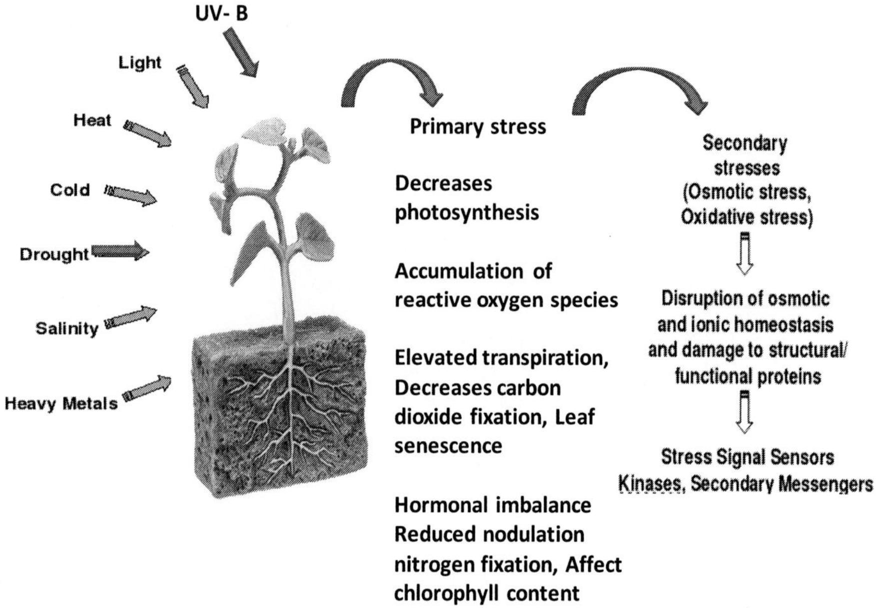

28. Responses of plants to various abiotic stresses.

Some investigations into combined UV-B and drought stress have shown it to reduce plant growth, photosynthetic capacity, pigment contents, biomass, and yield (Feng et al., 2007). In field-grown soybean, a decrease in productivity following UV-B exposure was shown to be moderated by soil water deficit (Sullivan and Teramura, 1990). It was found that radish seedlings were less sensitive to UV-B under water stress than cucumber seedlings. Radish showed higher leaf flavonoid contents, which possibly protected the seedlings by absorbing UV-B in the leaf epidermis. Nevertheless, the interaction between soil water deficit and UV-B stress in cowpeas resulted in benefits from the combined stresses in terms of greater growth and development compared to single stress exposure (Balakumar et al., 1993). Exposure to both UV-B and water stress was found to lead to decreased growth in cucumbers and radish, but protein content was increased. It was concluded that the UV-B irradiation could alleviate the negative effects of water stress on plants or exert an additional inhibitory effect on functional processes in soybeans (Tevini et al., 1983; Teramura et al., 1990). In contrast to Douglas fir seedlings grown in near-ambient UV-B radiation, decreased leaf area and increased leaf thickness were reported in seedlings in response to high UV-B radiation (Nagel et al., 1998). Both UV-B and drought stress were shown to have negative effects on the growth of *Arabidopsis thaliana* (Rajabbeigi et al., 2013). Nikolopoulos et al. (1995) reported that enhanced UV-B radiation may be beneficial for Mediterranean pines, alleviating the adverse effects of summer drought through restriction of cuticular transpiration. The effectiveness of UV-B radiation may be reduced in the presence of drought, resulting in growth delay or a concurrent increase in UV-B protective mechanisms. Both UV-B and drought stress radiation treatments resulted in lower shoot dry matter per plant, but there was no significant interaction between the two treatments. Zhang et al. (2011) also found that conditions of moderate drought stress and enhanced UV-B radiation increased biomass accumulation in the root and stem. Under conditions of severe drought stress, enhanced UV-B radiation led to increased biomass accumulation in the root, as well as some increase in biomass accumulation in the stem. This verifies that enhanced UV-B radiation can reduce the influence of biomass accumulation of *Fagopyrum dibotrys* under drought. The interactive effect of UV-B and drought stress on growth during recovery also has important consequences for the maintenance of reproductive success. Furthermore, reduced final yields of siliques were observed in plants subjected to combined stress treatment. Sangtarash et al. (2009) found combined UV-B and drought stress to have an additive effect in terms of the reduction of leaf area and biomass production in *Stellaria longipes*. One currently under-researched aspect of plant responses to both UV-B and drought is their role as regulators of plant phenology and reproductive success. The maintenance of reproduction following stress is often accompanied by changes in phenology (Pigliucci and Schlichting 1995; Brun et al. 2003). Feng et al. (2007) found UV-B and drought to have opposite effects on phenological timings in *Triticum aestivum*; therefore, UV-B and drought stress may have interactive effects on plant phenology. Conversely, Alexieva et al. (2001) found that *Pisum sativum* and *Triticum aestivum* exposed to combined UV-B and drought stress achieved greater total biomass than plants exposed to UV-B stress alone.

Physiological and biochemical responses to UV-B and drought

Global attention has become focused on increasing UV-B intensity and drought (due to low rainfall), which can have negative impacts on ecological and biological systems (Caldwell et al., 2007; Ballare et al.,

2011). Most of the stress combinations studied, including combined UV-B and drought stress, have been shown to have harmful effects on crops (Bandurska et al., 2013). It has been found that the combined effect of UV-B radiation and drought stress leads to alterations and reductions in plant growth, physiology, and biochemical processes (Alexieva et al., 2001). Cechin et al. (2008) found that sunflowers subjected to UV-B and drought stresses saw lower shoot dry matter per plant, as well as reduced photosynthesis, stomatal conductance, transpiration, and chlorophyll-a pigment. On the contrary, in white clover seedlings both stresses were demonstrated as having no significant effect on photosynthesis and biomass accumulation (Hofmann et al., 2003). Drought stressed cucumber plants under increased UV-B radiation lost their stomatal closing capacity at midday, saw reduced stomatal conductance and transpiration, and experienced additional negative effects on net photosynthesis (Tevini et al., 1983; Teramura et al. 1984; Kyparissis et al., 2001; Feng et al., 2007).

At the physiological level, other factors potentially occur in response to restrictive UV-B effectiveness in drought stressed plants. For example, it has been found that exposure to high UV-B radiation alongside drought may reduce sensitivity and cause plant phosphorus deficiency. In addition, a reduction in growth, altered intercellular CO_2 concentration, and apparent limitations on the assimilation of photosynthetic capacity in soybean were found in response to both UV-B and drought stress. However, the amplitude of the effects of both stressors was dependent on their interactions. Water stress reduced photosynthetic pigments only under high UV-B radiation and the decrease was more accentuated for chlorophyll-a than for chlorophyll-b (Sullivan and Teramura, 1990; Feng et al., 2007). An adverse effect in response to UV-B radiation and drought stress in wheat was observed by Tian and Lei (2007). Hui et al. (2016) found that the combined application of UV-B and water-deficit produced significantly higher chlorophyll, carotenoid, and total flavonoid contents in *Bryum argenteum* plants. These results suggest that drought stress alleviates the negative effects caused by enhanced UV-B radiation.

The interactive effects of UV-B and drought stress on the production of secondary metabolites do not show a clear pattern. Prior treatment with a high ratio of UV-B to PAR treatment enhanced the production of flavonoids in pea plants that were subsequently subjected to drought conditions (Nogués et al., 1998). Conversely, simultaneous UV-B treatment and drought stress dramatically reduced the UV-B mediated induction of anthocyanins and flavonols in barley (Bandurska et al., 2012) and pea plants (Alexieva et al., 2001). When combined with drought, Alonso et al.

(2015) reported lower UV-B mediated induction of terpene compounds in grapevines (*Vitis vinifera*). Conversely, Llusia et al. (2012) showed that terpene emissions were altered by increased UV-B and drought stress in a species-specific manner in Mediterranean species of xerophytes (*Daphne gnidium* and *Pistacia lentiscus*) and mesophytes (*Ilex aquifolium* and *Laurus nobilis*). In one of these species, combined drought stress and UV-B conditions elicited a stronger response (Nguyen et al., 2016; Bravo et al., 2017). Hassan et al. (2013) observed the highest levels of anthocyanin and flavonoids after exposure to UV-B radiation, while the lowest content was found in drought induced *Vicia feba*. In addition, higher proline content in *Arabidopsis* plants under enhanced UV-B radiation may be responsible for its higher tolerance to drought stress (Poulson et al., 2006; Cechin et al., 2008).

Previous investigations of UV-B and drought stress have also indicated that accumulation of osmolytes is an important determinant of interaction between these factors. For example, in cases where UV-B has been shown to reduce the negative impact of concurrent drought stress, it is the accumulation of low molecular weight and soluble metabolites, such as sugars, which have been implicated (Schmidt et al., 2000; Alexieva et al., 2001). Drought has previously been shown to reduce the negative impact of UV-B on yield in *Glycine max* (Sullivan and Teramura, 1990). UV-B as well as drought is known to promote the activity of enzymes playing an important role in phenol metabolism, such as phenylalanine ammonia-lyase (PAL) (Dixon and Paiva, 1995; Treutter, 2010). Stress mediated PAL synthesis induced by UV-B has been documented in lettuce (Caldwell and Britz, 2006) and white asparagus spears (Eichholzet al., 2012). The amino acid proline is a prerequisite marker of drought stress (Alexievaet al., 2001) and may also act as a protective factor against UV stress (He et al., 2011). Furthermore, the impact of combined drought and UV-B radiation led to much higher total anthocyanin. It was found in a number of studies that PAL activity increased in response to stress impacts like UV and drought stress (Oh et al., 2010; Shehab et al., 2010; Rajabbeigi et al., 2013). Similarly, Tian and Lei (2007) found combined factors to have an additive negative effect on the growth of *Triticum aestivum* seedlings, possibly related to enhanced oxidative damage. Similarly, He et al. (2011) found that the pre-application of drought caused increased tolerance to UV-B and vice versa in *T. aestivum*, suggesting some cross-tolerance between these factors. Kilian et al. (2007) identified a substantial overlap in gene expression in *A. thaliana* in response to both UV-B and drought. Similarly, Schmidt et al. (2000) identified ameliorative effects of combined UV-B and drought stress on the maintenance of leaf water content in *A. thaliana*. Nevertheless, whilst drought and UVB have been shown to interact, the responses found have not always been consistent between studies.

Effect of UV-B and drought on oxidative stress and the antioxidant system

The combined action of water deficit and UV-B radiation can modify plant response patterns. They can interact in an additive or synergistic manner increasing stress, but can also interact to greatly increase plant resistance. The amplification of negative effects may occur under the influence of simultaneous, as well as sequential action of prolonged or severe stress. However, there is only limited data in the literature relating to the interplay between drought and UV-B applied sequentially, as often occurs in nature. Many contradictory results about antioxidant enzyme responses to different stresses have emerged due to the fact that the levels of enzyme responses depend on the plant species, the plant's developmental stage, and the organs involved, as well as on the duration and severity of the stress (Rout and Shaw, 2001). In many plants, free proline accumulates in response to biotic and abiotic stresses, including UV-B irradiation (Carletti et al., 2003).

Tian and Lei (2007) concluded that drought stress and UV-B irradiation can both cause oxidative damage to plants through the generation of excessive reactive oxygen species (ROS). UV-B was shown to cause more severe stress than drought stress and the effect of UV-B and drought stress was seen to be additive in wheat seedlings. However, it is known that an excess of UV-B radiation and drought individually leads to accelerated production of ROS, such as singlet oxygen ($\cdot O_2$),superoxides (O_2^-), hydrogen peroxide (H_2O_2), and hydroxyl radicals ($OH\cdot$). Free radicals are extremely reactive and directly oxidize many biomolecules, such as the phospholipids of plasma membrane, nucleic acids, and proteins. Moreover, they can lead to the modification and destruction of proteins and enzymes in polymers, such as ribulose-1,5-bisphosphate carboxylase (RUBISCO), ATPase violaxanthin de-epoxidase, and the protein subunit of photosystems I and II (Rai and Agrawal, 2017) leading to irreparable metabolic dysfunction and death of plant cells. On the other hand, some plants have well-developed resistance mechanisms for stress tolerance to UV-B radiation and drought stress in the form of an antioxidant system. Non-enzymatic antioxidants also play a role, including ascorbic acid, polyamines, tocopherol, carotenoids, alkaloids, and flavonoids (Rao et al., 1996);

antioxidant enzymes, such as superoxide dismutase (SOD), ascorbate peroxidase (APX), catalase (CAT), monodehydroascorbate reductase (MDHAR), dehydroascorbate reductase (DHAR), glutathione S-transferase (GST), glutathione reductase (GR), glutathione peroxidase (GPX), and peroxidases (POX); and non-enzymatic compounds, such as ascorbate (ASA), glutathione (GSH) (Ueda and Nakamura, 2011; Siddiqui et al., 2011; Mackemess et al., 2001; Fedina et al. 2009; Broshe and Strid, 2003). Antioxidative responses of plants towards UV and drought stress are presented in [**29**].

Equally, UV-B and drought stress were seen to cause the formation of ROS; induce the synthesis of plant hormones, like absicic acid (ABA), ethylene, JA (Jasmonic acid), and SA (salicylic acid); activate defense mechanisms, including the accumulation of flavonoids (UV-B absorbing), polyamines, LEA proteins, and dehydrins; and improve the leaf water relationship through stomatal closure, osmolyte accumulation, and enzymatic and non-enzymatic antioxidant system upregulation (Siddiqui, 2011; Mackerness, 2001). On the contrary, prolonged action or high acute doses are potentially cytotoxic due to the generation of ROS, which can directly harm the normal function and components of cells (Sharma et al., 2012). Cellular damage and plant survival activity can be determined by the relation of pre-oxidants and antioxidant activity. The combined action of UV-B stress and drought stress is of great detriment to plants. UV-B radiation's combined action has an additive detrimental effect in wheat seedlings; when acting alone, it causes oxidative injuries and growth retardation (Tian et al., 2007).

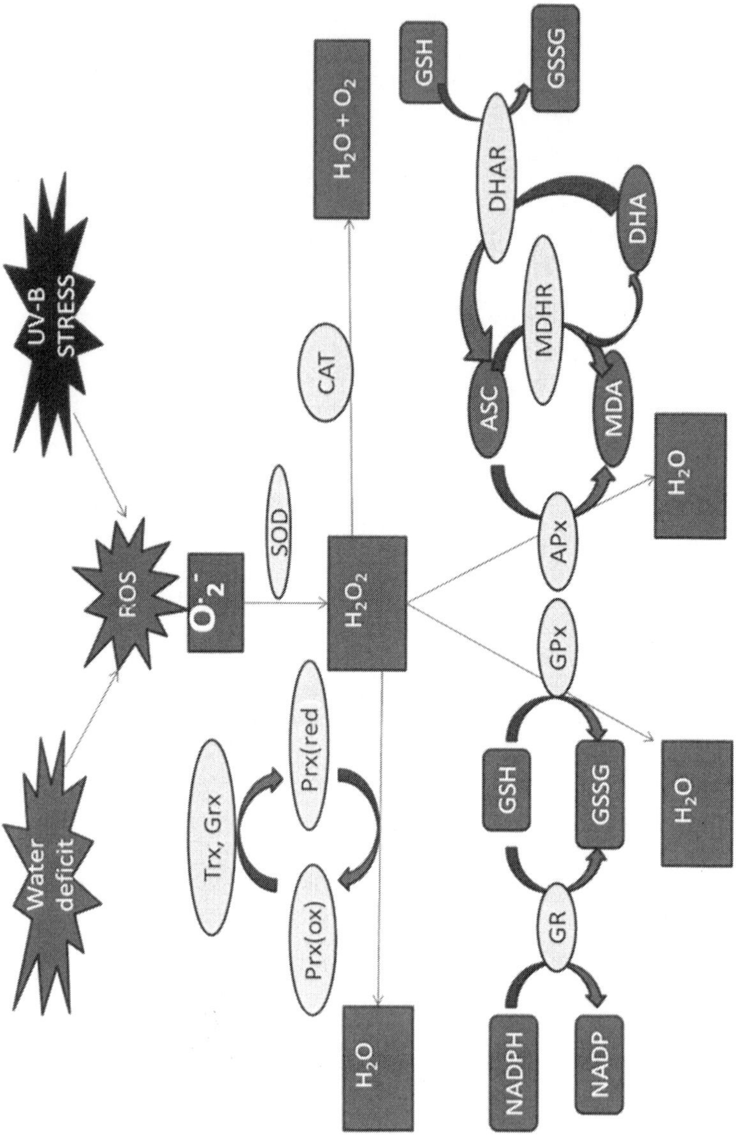

29. Antioxidative system in response to UV-B and drought stress.

Under natural field conditions, drought stress and high UV-B radiation operate separately; frequently, however, plants experience both stresses simultaneously. The co-occurrence of these stresses can alter the effect of individual stress responses. UV-B stress was shown to be alleviated by simultaneously applied moderate drought stress conditions, which led to an increase in catalase activity and flavonoid levels, thus reducing oxidative damage caused by UV-B in wheat and pea plants (Alexieva et al., 2001; Feng et al., 2007). Pre-treatment of plants with mild drought stress prior to UV-B radiation effectively reduced the level of H_2O_2, prevented lipid peroxidation, and reduced the rate of photosynthesis in tobacco (Hideg et al., 2003). Drought stress induced enhanced accumulation of UV-B absorbing compounds, caused a decrease in sensitivity to biologically effective UV-B radiation in *Trifolium repens*, and reversed the negative effect of UV-B radiation on growth in *Populus yunnanensis* (Hofmann et al., 2003; Daun et al. 2008). Tian and Lei (2007) found that the combined action of both stressors retarded seedling growth and increased the antioxidant activity of enzymes like superoxide dismutase, guaiacol peroxidase, and ascorbate peroxidase in wheat seedlings. Similar results have been noted in *Vigna unguiculata* L. (Balakumar et al., 1993), sunflowers (Cechin et al., 2008), *Picea asperata* (Lu et al., 2007), and barley seedlings (Bandurska et al., 2012).

Balouhci et al. (2009) found an increase in the proline contents of bean leaves in response to UV-B radiation and drought stress. The combined action of drought stress and UV-B radiation can modify response patterns in plants. For example, Siddiqui (2011) reported that simultaneous exposure to UV-B and drought stress causes the formation of ROS, H_2O_2, and NO; induces the synthesis of plant hormones like ABA, ethylene, jasmonic acid, and salicylic acid; and activates defense mechanisms, such as the accumulation of UV-B absorbing flavonoids, polyamines, and dehydrins etc.

When UV-B causes damage, the combined action of drought and UV-B usually becomes more detrimental. The combined action of moderate drought and ultra-high dose UV-B radiation had an additive detrimental effect in wheat seedlings (Tian and Lei, 2007). In barley seedlings the combined application of these stresses reduced the positive effect of UV-B on the synthesis of flavonoids, leading to higher membrane injury (Bandurska et al., 2012). Combined stress treatment also led to inhibition of the accumulation of soluble sugars, restricting the capacity for osmotic adjustment and having a negative additive effect on rosette growth (Comont et al., 2012). The negative effects of high ambient UV-B were shown to be alleviated by the simultaneous application of moderate drought conditions, leading to an increase in catalase activity and flavonoid levels, reducing the oxidative damage caused by UV-B in wheat and pea plants (Alexieva et al., 2001; Fang et al., 2007). Pre-treatment of plants with mild drought prior to UV-B effectively reduced the level of H_2O_2, preventing lipid peroxidation, and a decrease in the photosynthetic rate of tobacco exposed to high-dose UVB. This improved tolerance to UV-B stress was primarily derived from the over expression of aldose/aldehyde reductase engages in the detoxification of lipid peroxidation products (Hideg et al., 2003).

It should be noted that drought induced accumulation of UV-B absorbing compounds in *P. yunnanensis* was accompanied by an increase in ABA concentration. Moreover, exogenous application of ABA was shown to significantly increase the accumulation of these compounds (Tossi et al., 2011, 2012a, 2012b). In some plants, a water deficit was shown to induce SA acid accumulation (Bandurska, 2013; Bandurska and clislak, 2012; Banduraska and stroinski, 2005). SA may play a role in water deficit induced mechanisms of cross-resistance to UV-B. Exogenous application of SA attenuated UV-B induced damage by increasing the level of superoxide dismutase and catalase activity, as well increasing the concentration of anthocyanins, leading to improved growth under UV-B stress (Pael et al., 2013). Taking into account the rapid increase in JA concentration in water stressed plants, it can be assumed that also this hormone regulates water deficit induced resistance to UV-B (Wastemack, 2007).

Supplemental UV-B radiation was observed to cause a two-fold increase in cuticule thickness, enabling the plant to avoid excessive water loss, improve water economy, and alleviate the adverse effects of summer drought on the efficiency of the photosynthetic apparatus (Manetas et al., 1997; Petropoulou et al., 1995). Similarly, seedlings of *Pseudotsuga menziesii* grown under high UV-B were more resistant to drought because they saw lower transpirational water losses and maintained better water status than seedlings grown without UV-B radiation (Paulson et al., 2002). Schmidt et al. (2000) found that UV-B induced improvment of leaf water status may be associated with the production of osmolytes and the accumulation of dehydrins, but not with stomatal closure. Moreover, reduced oxidative damage was observed in barley leaf cell membranes (Bandusrla and Clislak, 2012). The results of a number of experiments have revealed that the simultaneous application of UV-B radiation with water deficit compensated the negative effects of moderate drought on the activity of antioxidant enzymes (superoxide dismutase, catalase) in *Vigna unguiculata* L. (Balakumar et al., 1993) and those of severe drought on lipid peroxidation in sunflowers (Cechin et al., 2008). The impact of UV-B radiation on alleviating drought induced cell membrane damage through a reduction in electrolyte leakage and a decrease in the level of lipid peroxidation was also found in *Picea asperata* populations (Lu et al., 2007).

Taking into account the increase in SA concentration in barley seedlings after pre-treatment with supra-physiological UV-B and the likely role of salicylate in plant responses to water deficits, we can assume that it is involved in UV-B induced cross resistance to drought (Bandurska and Clislak, 2012; Saruhan el al., 2012). The protective action of SA may include improvements in leaf water status through osmotic adjustment and stomatal closure, as well as the alleviation of water deficit induced injury by upregulation of the antioxidant system (Bandurska et al., 2013). Bearing in mind that UV-B causes an increase in ABA and NO concentrations, these regulators may have a role in mediating the UV-B induced alleviation of oxidative damage in water deficit stressed plants (Tossi et al., 2012). It has been postulated that ABA induced activation of antioxidant enzymes involve H_2O_2 dependent NO generation, which, in turn, activates mitogen-activated protein kinase cascades and gene expression (Zhang et al., 2007). However, ethylene and JA, whose concentrations increase on exposure to UV-B, may support ABA dependent defence mechanisms, or they may operate independently to increase water deficit resistance. It has been shown that ethylene and JA interact with ABA mediated stomatal closure via H_2O_2 generation and enhanced production of NO (Acharya and Assmann, 2009; He et al., 2011).

Conclusions and future prospects

The combined effect of UV-B and drought stress is of increasing interest in relation to the growth, development, physiology, biochemical changes, and antioxidant activity of plants. The elevation of both stresses at the same time has been found to cause changes in plant growth, physiology, biochemistry, and oxidative stress activity. Several components of the photosynthetic apparatus are affected by these stress factors at the molecular level. Alongside this, there have been significant advances in our understanding of the combined effects of these factors on terrestrial ecosystems, particularly in terms of plant response mechanisms.

Understanding the many underlying developmental and resistance mechanisms in plants has allowed us to observe their mediatiation of UV-B and drought stress. As a result, we can better understand common responses to these combined stress factors in plants, such as diminished growth, acclimation, and the stimulation and transmission of signals within the plants. The resulting modifications in gene expression may also alter common physiological, biochemical, and antioxidant responses in plants, or they may generate overlapping functions. However, the combined effect of UV-B and drought stress in plants is still not well understood and more research needs to be carried out under field conditions to provide definitive answers in this area.

References

Acharya, BR, and Assmann, SM. 2009. "Hormone interactions in stomatal function". *Plant Molecular Biology*, No. 69: 45-462.

Alexieva, V, Sergiev, I, Mapelli, S, Karanov, E. 2001. "The effect of drought and ultraviolet radiation on growth and stress markers in pea and wheat". *Plant Cell Environment*, No. 24:1337-1344.

Alonso R, Berli FJ, Bottini R, Piccoli P. 2015. "Acclimation mechanisms elicited by sprayed abscisic acid, solar UV-B and water deficit in leaf tissues of field-grown grapevines". *Plant Physiology Biochemistry*, No. 91: 56-60.

Balakumar, T, Vincent, VHB, Paliwal, K. 1993. "On the interaction of UV-B radiation (280-315 nm) with water-stress in crop plants". *Physiology of Plants*, No. 87: 217-222.

Ballaré, CL, Caldwell, MM, Flint, SD, Robinson, SA, Bornman, JF. 2011. "Effects of solar ultraviolet radiation on terrestrial ecosystems. Patterns, mechanisms, and interactions with climate change". *Photochemical and Photobiological Science*, No. 10: 226-241.

Balouchi, HR, Sanavy, SAM, Emam, Y, and Dolatabadian, A. 2009. "UV radiation, elevated CO_2 and water stress effect on growth and photosynthetic characteristics in durum wheat". *Plant Soil Environment*, No. 55: 443-453.

Bandurska, H, and Cieślak, M. 2013. "The interactive effect of water deficit and UV-B radiation on salicylic acid accumulation in barley roots and leaves". *Environmental and Experimental Botany*, No. 94: 9-18.

Bandurska, H, and Stroiński, A. 2005. "The effect of salicylic acid on barley response to water deficit", *Acta Physiologiae Plantarum*, No. 27: 379-386.

Bandurska, H, Hayat, A, Ahmad, A, and Alyemeni, MN. 2013. "Salicylic acid: an update on biosynthesis and action in plant response to water deficit and performance under drought". *Plant Growth and Development*, 1-14.

Bandurska, H, Pietrowska-Borek, M, and Cieślak, M. 2012. "Response of barley seedlings to water deficit and enhanced UV-B irradiation acting alone and in combination". *Acta Physiologiae Plantarum*, No. 34: 161-171.

Bravo, ER, Klinkhamer, PG, and Leiss, KA. 2017b. "Interactive effects of UV-B light with abiotic factors on plant growth and chemistry, and their consequences for defense against arthropod herbivores". *Frontiers in Plant Science*, No. 8: 278.

Bravo, ER, Klinkhamer, PGL, and Leiss, KA. 2017a. "Induction of jasmonic acid-associated defenses by thrips alters host suitability for conspecifics and correlates with increased trichome densities in tomato". *Plant & Cell Physiology*, No. 58: 622-634.

Brosche, M, and Strid, A. 2003. "Molecular Events Following Perception of Ultraviolet-B Radiation by Plants". *Physiology of Plant*, No. 117: 1-10.

Brun, LA, Maillet, J, and Le Corff J. 2003. "Effects of elevated soil copper on phenology, growth and reproduction of five ruderal plant species". *Environment and Pollution*, No. 122: 361-368.

Caldwell, CR, and Britz, S. 2006. "Effect of supplemental ultraviolet radiation on the carotenoid and chlorophyll composition of green housegrown leaf lettuce (*Lactuca sativa* L.) cultivars". *Journal of Food Composition and Analysis*, No. 19: 637-644.

Caldwell, MM, Bornman, J, Ballaré, C, Flint, SD, and Kulandaivelu, G. 2007. "Terrestrial ecosystems, increased solar ultraviolet radiation, and interactions with other climate change factors". *Photochemical and Photobiological Science*, No. 6: 252-266.

Carletti, P, Masi, A, Wonisch, A, Grill, D, Tausz, M, and Ferretti, M. 2003. "Changes in antioxidant and pigment pool dimensions in UV-B irradiated maize seedlings". *Environmental and Experimental Botany*, No. 50: 149-15.

Cechin, I, Corniani, N, Fumis, TDF, and Cataneo, AC. 2008. "Ultraviolet-B and water stress effects on growth, gas exchange and oxidative stress in sunflower plants". *Radiation and Environmental Biophysics*, No. 47: 405-413.

Comont, D, Winters, A, and Gwynn-Jones, D. 2012. "Acclimation and interaction between drought and elevated UV-B in *A. thaliana*:

differences in response over treatment, recovery and reproduction". *Ecology and Evolution*, No. 2: 2695-2709.

Cushman, JC, and Bohnert, J. 2000. "Genomic approaches to plant stress tolerance". *Current Opinion in Plant Biology*, No. 3(2): 117-124.

Dixon, RA and Paiva, NL 1995. "Stress-induced phenylpropanoid metabolism". Plant Cell, No. 7: 1085-1097.

Duan, BL, Xuan, ZY, Zhang, XL, Korpelainen, H, and Li, CY. 2008. "Interactions between drought, ABA application and supplemental UVB in *Populus yunnanensis*". *Physiologia Plantarum*, No. 134: 257-269.

Eichholz, I, Gamm, A, Herppich, WB, Ulrichs, C, Rohn and Huysken-kill, S. 2012. "UV-B mediated changes in enzyme activity inducing flavonoids of white asparagus spears". *Food Research and International*, No. 48: 196-201.

Fang, CX, Wang, QS, Yu, Y, Huang, LK, Wu, XC, and Lin, WX. 2011. "Silicon and its uptaking gene Lsi1 in regulation of rice UV-B tolerance". *Acta Agronomica. Sinica*, No. 37(06): 1005-1011.

Fedina, I, Nedeva, D, Georgieva, K, and Velitchkova, M. 2009. "Methyl jasmonate counteracts UV-B stress in barley seedlings". *Journal of Agronomy and Crop Science*, No. 195: 204-212.

Hassan, IA, Basahi, JM, Haiba, NS, and Kadi, MW. 2013. "Investigation of Climate Changes on Metabolic Response of Plants: Interactive Effects of Drought Stress and Excess UV-B". *Earth Science and Climate Change*, No. 4(1): 129.

He, J, Yue, X, Wang, R, and Zhang, Y. 2011. "Ethylene mediates UV-B-induced stomatal closure via peroxidase-dependent hydrogen peroxide synthesis in *Vicia faba* L.". *Experimental Botany*, No. 62: 2657-26.

Hide, E, Nagy, T, Oberschall, A, Dudits, D, Vass, I. 2003. "Detoxification function of aldose/aldehyde reductase during drought and ultraviolet-B (280–320) stresses". *Plant Cell Environment*, No. 26: 513-522.

Bandurska, H, Borek, MP, and Cieślak, M. 2003. "Responses to UV-B radiation in *Trifolium repens* L.: physiological links to plant productivity and water availability". *Plant Cell Environment*, No. 26: 603-612.

Hui, R, Zhao, R, Liu, L, Zhu, R, Li, G, and Wei, Y. 2016. "Effects of UV-B, water deficit and their combination on *Bryum argenteum* plant". *Russian Journal of Plant Physiology*, No. 63: 216-223.

Kanojia, A, and Dijkwel, P. 2018. "Abiotic Stress Responses are Governed by Reactive Oxygen Species and Age". *Annual Plant Reviews*, 1-32.

Kataria, S, and Guruprasad, KN. 2014. "Exclusion of solar UV components improves growth and performance of *Amaranthus tricolor* varieties". *Scientia Horticulturae*, No. 174: 36-45.

Kilian, J, Whitehead, D, Horak, J, Wanke, D, Weinl, S, Batistic, O. 2007. "The global stress expression data set: protocols, evaluation and model data analysis of UV-B light, drought and cold stress responses". *Plant Journal*, No. 50: 347-363.

Kyparissis, A, Drilias, P, Petropoulou, Y, Grammatikopoulos, G, and Manetas, Y. 2001. "Effects of UV-B radiation and additional irrigation on the Mediterranean evergreen sclerophyll *Ceratonia siliqua* L. under field conditions". *Plant Ecology*, No. 154: 187-193.

Llusia, J, Llorens, L, Bernal, M, Verdaguer, D, and Penuelas, J. 2012. "Effects of UV radiation and water limitation on the volatile terpene emission rates, photosynthesis rates, and stomatal conductance in four Mediterranean species". *Acta Physiology of Plant*, No. 34: 757-769.

Lu, Y, Duan, B, and Li Ch. 2007. "Physiological responses to water deficit and enhanced UV-B radiation in two contrasting *Picea asperata* population". *Canadian Journal of Research*, No. 37: 1253-1262.

Mackerness, AH, John, CF, Jordan, B, and Thomas, B. 2001. "Early signalling components in ultraviolet-B responses: distinct roles for different reactive oxygen species and nitric oxide". *Federation of European Biochemical Societies*, No. 489: 237-242.

Manetas, Y, Petropoulou, Y, Stamatakis, K, Nikolopoulos, D, Levizou, E, Psaras, G, and Karaboumlotis, G. 1997. "Beneficial effects of enhanced UV-B radiation under field conditions: Improvement of needle water-relations and survival capacity of *Pinus pinea* L. seedlings during the dry Mediterranean summer". *Plant Ecology*, 128: 100-108.

Mittler, R. 2006. "Abiotic stress, the field environment and stress combination". *Trends Plant Science*, No. 11: 15-19.

Mittler, R, Merquiol, E, Hallak-Herr, E, Rachmilevitch, S, Kaplan, A, and Cohen, M. 2001. "Living under a 'dormant' canopy: a molecular acclimation mechanism of the desert plant *Retama raetam*". *Plant Journal*, No. 25: 407-416.

Nagel, LM, Bassman, JH, Edwards, GE, Robberecht, R, and Franceshi, VR. 1998. "Leaf anatomical changes in *Populus trichocarpa*, *Quercus rubra*, *Pseudotsuga menziesii* and *Pinus ponderosa* exposed to enhanced ultraviolet-B radiation". *Physiology of Plant*, No. 104: 385-396.

Nikolopoulos, D, Petropoulou, Y, Kyparissis, A, and Manetas, Y. 1995. "Effects of enhanced UV-B radiation on the drought semideciduous

Mediterranean shrub *Phlomis fruticosa* under field conditions are season-specific". *Australian Journal of Plant Physiology*, No. 22: 737-745.

Nogues, S, Allen, DJ, Morison, JJL, and Baker, NR. 1998. "Ultraviolet-B radiation effects on water relations, leaf development, and photosynthesis in water deficited pea plants". *Plant Physiology*, No. 117: 173-181.

Oh, MM, Carey, EE, and Rajashekar, CB. 2010. "Regulated water deficits improve phytochemical concentration in Lettuce". *American Society of Horticulture Science*, No. 135: 223-229.

Pael, M, Hayat, A, Ahmad, A, and Alyemeni, MN. 2013. *Salicylic Acid. Plant Growth and Development*. Springer: New York/London. 163-182.

Petropoulou, Y, Kyparissis, A, Nikolopoulos, D, and Manetas, Y. 1995. "Enhanced UV-B radiation alleviates the adverse-effects of summer water deficit in 2 Mediterranean pines under field conditions." *Physiology of Plants*, No. 94: 37-44.

Pigliucci, M, Schlichting, CD. 1995. "Reaction norms of *Arabidopsis Brassicaceae* III. Response to nutrients in 26 populations from a worldwide collection". *Am. J. Bot.*, No. 82: 1117-1125.

Pinheiro, C, and Chaves, MM. 2011. "Photosynthesis and drought: can we make metabolic connections from available data?". *Experimental Botany*, No. 62(3): 869-882.

Poulson et al. 2002. "Enhanced tolerance of photosynthesis to high-light and water deficit in *Pseudotsuga manziensii* seedlings grown in ultraviolet-B radiation". *Tree Physiology*, No. 22: 829-838.

Poulson, ME, Boeger, MRT, and Donahue, RA. 2006. "Response of photosynthesis to high light and drought for *Arabidopsis thaliana* grown under a UV-B enhanced light regime". *Photosynthesis Reseach*, No. 90: 79-90.

Rai, K, and Agrawal, SB. 2017. "Effects of uv-b radiation on morphological, physiological and biochemical aspects of plants: an overview". *Journal of Scientific Research*, No. 61: 87-113.

Rao, MV, Paliyath, G, Ormrod, DP. 1996. "Ultraviolet-B and Ozone-Induced biochemical changes in antioxidant enzymes of *Arabidopsis thaliana*". *Plant Physiology*, No. 110: 125-136.

Rajabbeigi, E, Eichholz, I, Beesk, N, Ulrichs, C, Kroh, LW, and Rohn, S. 2013. "Interaction of drought stress and UV-B radiation: impact on biomass production and flavonoid metabolism in lettuce (*Lactuca sativa* L.)". *J. Applied Botany and Food Quality*, No. 86: 190-197.

Rejeb, I, Pastor, V, and Mauch-Mani, B. 2014. "Plant responses to simultaneous biotic and abiotic stress: molecular mechanisms". *Plants*, No. 3: 458-475.

Rout, NP, and Shaw, BP. 2001. "Salt tolerance in aquatic macrophytes: possible involvement of the antioxidative enzymes". *Plant Science*, No. 160: 415-423.

Sangtarash, MH, Qaderi, MM, Chinnappa, CC, and Reid DM. 2009. "Differential responses of two *Stellaria longipes* ecotypes to ultraviolet-B radiation and drought stress". *Flora-Morphology Distribution, Functional Ecology of Plants*, No. 204: 593-603.

Saruhan, N, Saglam, A, and Kadioglu, A. 2012. "Salicylic acid pretreatment induces, water deficit tolerance and delays leaf rolling by inducing antioxidant systems in maize genotypes". *Acta Physiology of Plant*, No. 34: 97-106.

Schmidt, AM, Ormrod, DP, Livingston, NJ, and Misra, S. 2000. "The Interaction of Ultraviolet-B Radiation and Water Deficit in Two *Arabidopsis thaliana* Genotypes". *Annual Botany*, No. 85: 571-575.

Shao, H, Chu, L, Jaleel, C, and Zhao, C. 2008. "Water-deficit stress-induced anatomical changes in higher plants". *Comptes Rendus Biologies*, No. 331: 215-225.

Sharma, P, Jha, AB, Dubey, RS, and Pessarakl, M. 2012. "Reactive oxygen species, oxidative damage, and antioxidative defense mechanism in plants under stressful conditions". *Journal of botany*, 217037.101155/2012/217037.

Shehab, GG, Ahmed, OK, and El-Beltagi, HS. 2010. "Effects of various chemical agents for alleviation of drought stress in rice plants (Oryza sativa L.)". *Notulae Botanicae Horti Agrobotanici Cluj-Napoca*, No. 38: 139-148.

Siddiqui, MH, Al-Whaibi, MH, and Basalah, MO. 2011. "Role of nitric oxide in tolerance of plants to abiotic stress". *Protoplasma*, No. 248: 447-455.

Sullivan, JH, and Teramura, AH. 1990. "Field Study of the Interaction between Solar Ultraviolet-B Radiation and Drought on Photosynthesis and Growth in Soybean". *Plant Physiology*, No. 92: 141-146.

Teramura, AH, Foresth, IN, and Lydon, J. 1984. "Effects of UV-B Radiation on the Plants during Mild Water Stress: 4. The Insensitivity of Soybean Internal Water Relations to UV-B Radiation". *Physiology of Plants*, No. 62: 384-389.

Teramura, AH, Sullivan, JH, and Lydon, J. 1990. "Effects of UV-B Radiation on Soybean Yield and Seed Quality: A 6-Year Field Study". *Physiology of Plants*, No. 80: 5-11.

Tevini, M, and Iwanzik, W, and Teramura, AH. 1983. "Effects of UV-B Radiation on Plants during Mild Water Stress". *Zeitschrift-fur-pflanzenphysiologie*, No. 110: 459-467.

Tian, XR, and Lei, YB. 2007. "Physiological responses of wheat seedlings to water deficit and UV-B radiation. Effect of exogenous sodium nitroprusside application". *Russian Journal of Plant Physiology*, No. 54: 676-682.

Tossi, V, Amenta, M, Lamattina, L, and Cassia, R. 2011. "Nitric oxide enhances plant ultravilet-B protection up-regulating gene expression of the phenyl propanoid biosynthetic pathway". *Plant Cell Environment*, No. 34: 909-921.

Tossi, V, Cassia, R, Bruzzone, S, Zochi, E, and Lamattina, L. 2012. "ABA says NO to UV-B: a universal response". *Trends Plant Science*, No. 17: 510-517.

Tossi, V, Lombardo, C, Cassia, R, and Lamattina, L. 2012. "Nitric oxide and flavonoids are systemically induced by UV-B in maize leaves". *Plant Science*, No. 193/194: 103-109.

Treutter, D. 2010. "Managing phenol contents in crop plants by phytochemical farming and breeding-visions and constraints". *International journal of Molecular Science*, No. 11: 807-857.

Ueda, T, Nakamura, C. 2011. "Ultraviolet-defense mechanisms in higher plants". *Biotechnology and Biotechnological. Equipments*, 25 2177-2182.

Wasternack, C. 2007. "Jasmonates: an update on biosynthesis, signal transduction and action in plant stress response, growth and development". *Annual Botany*, 100: 681-697.

Zhang, A, Jiang, M, Zhang, J, Ding, H, Xu, S, Hu, X, and Tan, M. 2007. "Nitric oxide induced by hydrogen peroxide mediates abscisic acid-induced activation of the mitogen-activated protein kinase cascade involved in antioxidant defense on maize leaves". *New Phytologist*, No. 175(1): 36-50.

Zhang, X, Zou, Z, Gong, P, Zhang, J, Ziaf, K, Li, H, Xiao, F, and Ye, Z. 2011. "Over-expression of microRNA169 confers enhanced drought tolerance to tomato". *Biotechnology Letters*, No. 33: 403-409.

Zhu, JK. 2002. "Salt and drought stress signal transduction in plants". *Annual Review of Plant Biology*, No. 53: 247-273.

CHAPTER EIGHT

BIOTIC STRESS IN PLANTS: A METABOLOMICS PERSPECTIVE

S VIJAYAN, M THAMPI AND MS JISHA

Biotic stress covers an array of negative influences on plants caused by living organisms, such as bacteria, fungi, viruses, nematodes, and protozoans. Biotic stress impairs crop health and reduces yields. Biotic stress initiates a cascade of events in plants, leading to metabolic changes and thus to plants activating immune responses towards the stress. Metabolomics is an area of advanced scientific study, which focuses on a broad, quantitative evaluation of metabolites produced in biological systems. This advanced field of research can be used effectively to address biotic stress. The plant metabolome consists of a huge array of primary and secondary metabolites, hormones, and signaling molecules with low molecular weights. Metabolites play an indispensable role in the health system, biomass production, and total architecture of plants. When subjected to biotic stress, plants produce certain stress-response metabolites to adapt to the new condition. For this reason, accurate metabolite profiling is a breakthrough in the investigation of biotic stress tolerance in plants. Moreover, this method has a lot of potential for future development and is budget-friendly. This advanced field of study will allow us to select superior quality plants and breeding materials. In the present review, an attempt is made to discuss the techniques and significant advancements obtained in the metabolomics study of plants to improve disease resistance and yield.

Introduction

The global annual crop yield is massively affected by both abiotic and biotic stresses. Abiotic stresses are environmental stresses faced by plants during growth. They include drought, salinity, heat, waterlogging, cold, nutrient deficiency, and heavy metal toxicity. These factors have been extensively studied by scientists in the quest to develop new tolerant varieties (Kushalappa and Gunnaiah, 2013). Infection by pathogens is the other major threat faced by plants in the natural environment. Biotic stress (microorganisms, pests, and parasites) causes massive reductions in crop yields and leads to increased use of chemicals as a part of disease management and pest control (Liu et al., 2016). Such stress elevates plant metabolic pathways and metabolic homeostasis is disrupted.

This leads to a multigenic process known as acclimation in which the plant seeks to reattain metabolic homeostasis (Shulaev et al., 2008). This is achieved through a cascade of events. In the primary stage, the stress is sensed by the plant; the signal transduction pathways are then activated to produce a different set of proteins and hormones to achieve metabolic homeostasis. The compounds involved in this process can be broadly classified into three: 1. compounds involved in the acclimation process; 2. compounds that appear in cells due to altered plant metabolism; 3. signal transduction compounds or molecules (Mittler, 2006). It is important to understand a plant's metabolomics in order to understand its stress response mechanisms to biotic and abiotic stresses, which will also help in the development of resistant or tolerant crops.

The plant immune mechanism is well developed and multilayered with both first-line physical and chemical defense mechanisms. In addition to the first-line defense mechanisms, plants also possess specific defense patterns that are triggered only after stress is sensed (Balmer et al., 2013). For this reason, it is necessary to study the complete set of primary and secondary metabolites, hormones, and other signaling molecules to understand the exact defense mechanisms elicited by plants in response to biotic stress. During infestation by a pathogen, the production of phytochemicals, such as isoprenoids, phenylpropanoids, alkaloids, fatty acids, antimicrobial compounds, and polyketides, are stimulated. Analysis of these metabolic changes through the use of advanced tools is called metabolomics (Dawid and Hille, 2018).

Recent years have witnessed an emerging and fascinating approach to metabolomics in crop management and improvement. Metabolomic tools have been integrated with genomics, proteomics, and transcriptomics to study the plant metabolome and clearly understand the defense mechanisms used to tackle biotic stress. Advanced metabolomic tools help scientists to detect, assess, and evaluate all possible metabolites available and thus uncover the exact biochemical pathways involved in the process of acclimation (Kumar et al., 2017). In this chapter, we briefly describe the importance of metabolomics in biotic stress management. In addition, a comprehensive overview of the advanced tools and techniques employed in metabolomics to investigate plant metabolites is also provided.

Metabolomics: an emerging approach to biotic stress in plants

Metabolomics is a rapidly growing area of science and technology that contributes to crop improvement through its approach to plant biotic stress. Metabolite profiling, metabolic fingerprinting, and targeted analysis are the three major focuses of metabolomics. In addition to the conceptual framework, more recent integrated research in metabolomics has been used to study the stress defense mechanisms (Castro-moretti et al., 2020). The phenotype of an organism is closely related to its metabolites, thus metabolomics can be used in the identification of gene function and monitoring of the plant's metabolism in its response to biotic and abiotic stresses. As biotic and abiotic stresses have a significant role in reducing crop yields, metabolomics is used to probe for unique metabolites during the lifecycle of the plant. Under conditions of biotic stress, plants show variation in their physiological and biochemical processes thereby triggering immune sensors and providing resistance, including effector-triggered immunity (ETI), pattern-triggered immunity (PTI), and termed-pattern recognition receptors (PRRs) (Boutrot et al., 2017). A flow chart representing the general steps involved in metabolomics-assisted improvement of crops is shown in [**30**].

Metabolite profiling and fingerprinting

In metabolite profiling, a single sample is screened to discover a group or set of metabolites and generate a clear idea about the hundreds of compounds found and the various chemical classes to which they belong (sugars, organic acids, amino acids, aromatic amines, fatty acids, and sugar alcohols). Metabolite profiling is similar to other techniques, such as functional genomics and nuclear magnetic resonance spectroscopy (NMR), liquid chromatography-mass spectrometry (LC-MS), capillary electrophoresis–mass spectrometry (CE-MS), and Fourier transform infrared spectroscopy (FTIR), which are all advanced technologies employed to obtain metabolite profiles (Shulaev, 2006). However, metabolic fingerprinting does not precisely quantify or identify all the compounds present in the sample.

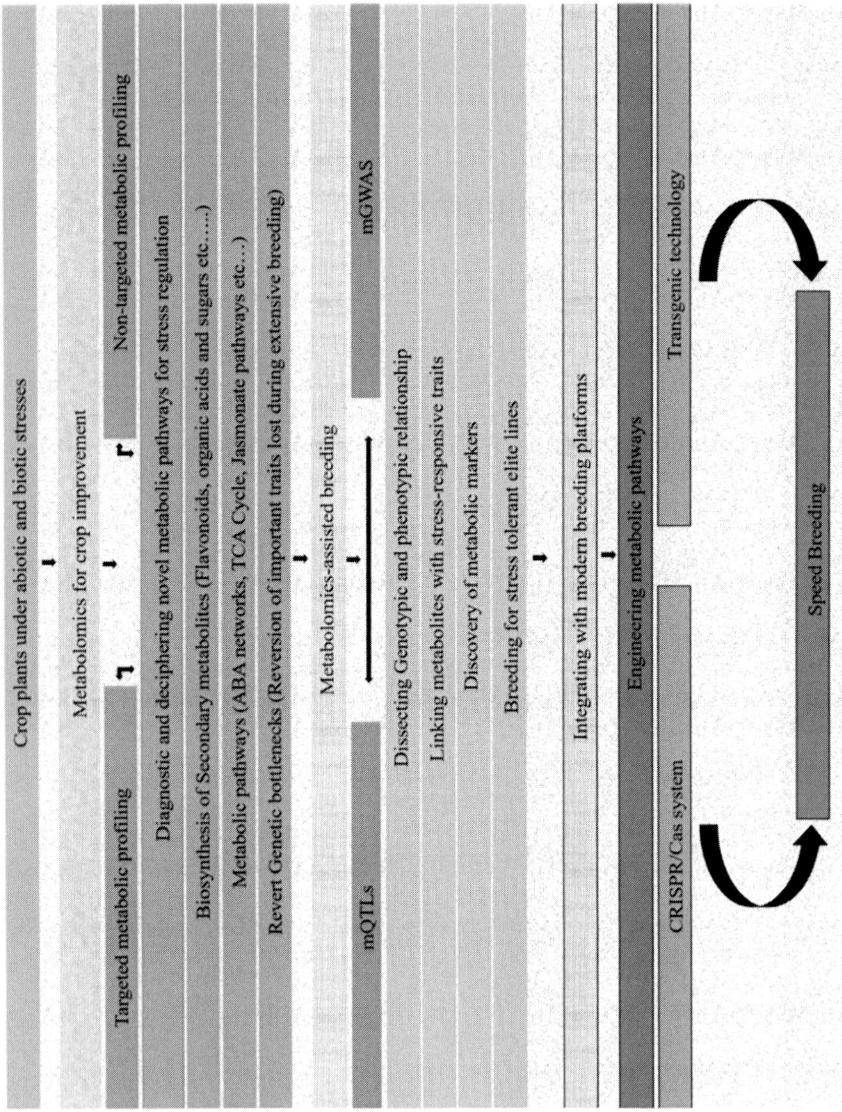

30. The plant metabolomics mechanism for crop improvement (Razzaq et al., 2019).

Rather, it focuses only on a limited profile or fingerprint from the sample and searches for the presence of unique metabolites in single tissues or cells. It helps to identify signature patterns that occur in certain cells in response to specific stress conditions (Krishnan et al., 2005). Metabolic fingerprinting can thus be employed to study the altered pattern of a plant's metabolism during biotic stress or infection by a pathogen. Pattern recognition and discrimination analysis can be combined with metabolic fingerprinting to analyze the data (Scholz et al., 2004; Castro-moretti et al., 2020). Comprehensive interpretation of metabolic fingerprinting data along with other omics data is relatively simple with the web-based tool MetaboAnalyst, which meets all the bioinformatic demands required (Chong et al., 2019).

Targeted analysis

This involves the quantitative analysis of target metabolites of a selected organism. Targeted analysis helps us understand plant metabolism under specific stress conditions. During biotic and abiotic stress, the metabolic pattern is altered and thus targeted analysis can provide relevant data on altered metabolic and signaling pathways (Ghosson et al., 2018). The major disadvantage associated with this analysis is that the analyst must already know the structure of the targeted metabolite. As it is a quantitative form of analysis, it is used in the analysis of metabolic change and not for the identification of novel marker metabolites (Shulaev, 2006; Roessner and Bowne, 2009).

The role of metabolomics in plant-pathogen interactions

Globally, crop loss due to biotic stress amounts to 20 to 30 % in major crops; this figure is even higher in food-insecure areas (Savary et al., 2019). To prevent crop loss from microbial or pathogen attack, researchers have developed pest resistant varieties through improved breeding techniques. Subsequently, they moved on to more advanced technologies that focus on such things as gene expressions and protein interactions. Model plant strains, like *Arabidopsis thaliana*, tomato, and maize, have all been used to study the metabolic aspects of plant-pathogen interaction in promising research. These studies have led researchers to recognize the importance of metabolites in plant-pathogen interactions. Finally, metabolomics has enabled a better understanding of biotic stress responses in plants (Solomon and Oliver, 2001; Zhou et al., 2011). Conventional plant disease studies were usually based on phenotypical comparison of symptoms between vulnerable and immune varieties of plants. Molecular studies on plant defense mechanisms associated with the cell wall, callose deposition at the site of infection, the production of reactive oxygen species, signaling molecules and hormone production, and the production of other antimicrobial agents, have revealed the importance of metabolites in plant defense mechanisms (Castro-moretti et al., 2020). Before metabolomics, other omics (genomics, transcriptomics, and proteomics) offered great support in detecting the immune response elicitation mechanisms of plants under biotic stress (Stahl et al., 1999; Stahl and Bishop, 2000; Bergelson et al., 2001).

Studies have revealed that *Fusarium* and mycotoxin resistance in cereals is due to the accumulation of certain secondary metabolites, including polypropanoides, lipophilic compounds, and benzoxazinoids during infection. This finding will help the development of resistant varieties (Atanasova-Penichon et al., 2016). Metabolomic analysis of resistant and susceptible soybeans infected with *Phytophthora sojae* revealed the accumulation of many sugars and secondary metabolites in a resistant variety compared to a susceptible one. This study emphasized how metabolic change occurs during infection and may lead to resistance or defense (Zhu et al., 2018). Pathogens produce effectors and toxins during infection to evade host defense mechanisms. Metabolomics helps us detect the influence of these molecules on the host's metabolism. *Fusarium* sp. is a major pathogen that infects cereals. Studies have revealed that they produce several toxins, including rubrofusarin, terpenes, and many secondary metabolites. All these products are responsible for making conditions favorable for pathogenic infection (Placinta et al., 1999). Atanasova-Penichon et al. (2016) described the metabolomic aspects of defense responses elicited by cereals against *Fusarium* sp. Their review identified metabolites that confer resistance in cereals to *Fusarium* infection. Flavonoids, non-flavonoid phenylpropanoids, benzoxazinoids, and lipophilic secondary metabolites are the major compounds mentioned in the review. These studies help highlight the importance of resistance mechanisms in cereals and the role of metabolomics and functional genomics in them (Atanasova-Penichon et al., 2016). A study of *Pseudomonas syringae* infection in resistant varieties of soybean revealed that the production of daidzein is induced during the infection. Daidzein is an immediate precursor of phytoalexins in soybeans, which act as antimicrobial agents and prevent the infection. In susceptible varieties, daidzein biosynthesis is inhibited. These findings from metabolomic analysis are useful in the development of resistant varieties (Zhou et al., 2011).

Studies on *Arabidopsis thaliana* disclosed that they produce some indolic secondary metabolites to defend against pathogens. They produce camalexin (a phytoalexin) and glucosinolate (a phytoanticipin) in response to the pathogen *Phytophthora brassicae*. The study was clarified further by analysis of mutant *Arabidopsis* plants, which fail to produce camalexin and glucosinolate and are more vulnerable to infection by *Phytophthora brassicae* (Schlaeppi and Mauch, 2010). An untargeted metabolomic approach towards the host-pathogen interaction in tomato plants infected with *Ralstonia solanacearum* showed that flavonoids and hydroxycinnamic acids play a primary role in the plants' defense responses. Infected tomato plants showed accumulation of feruloyl tyramine, feruloyl methoxytyramine, and caffeoyl putrescine in the tissues, which had a positive impact on pathogen defense (Zeiss et al., 2019). Furthermore, polyamines and spermidine were shown to accelerate the plant immune system in response to both abiotic and biotic stresses (Kasukabe et al., 2004; Hussain et al., 2011; Chen et al., 2019).

An increasing number of metabolites are being detected and identified as responsible for biotic stress tolerance. Gall midge resistant rice varieties produce about 16 fatty acids along with two amino acids (glutamine and phenylalanine) in response to infection and remain non-susceptible. Similar patterns of metabolic response have been observed in rice, barley, maize, and wheat to infection with *Magnaporthe grisea*. On infection, the production and accumulation of malate, polyamines, quinate, and non-polymerized lignin precursors were all detected. In wheat, the production and accumulation of phenylpropanoid and phenolic compounds were identified during infection with *Fusarium graminearum*. Phenylpropanoids are precursors of lignin, enhancing lignin production in plants to resist infection. As can be seen, metabolomics gives us a better understanding of the chemical machinery that operates in plants during biotic stress (Hussain et al., 2011; Hong et al., 2016).

The major obstacle faced by researchers into the metabolomics of plant-pathogen interactions is discrimination of pathogen and plant metabolites. A targeted approach in which microbial or pathogen metabolites are labeled with isotope tagging or labeling can help overcome this situation. Another difficulty faced in research is the identification of unknown compounds, as there is no proper database available for plant metabolites. Metabolomics is time consuming and requires expensive technology. Isolation, purification, and analysis of metabolites need highly qualified researchers with suitable skillsets. The integration of other omics with metabolomics is useful for better understanding and analysis of data, but it is not easy to perform (Pinu et al., 2019). Progress in multi-omics data integration and collaboration will empower researchers to construct usable databases and platforms. Amalgamating all the information will lead to a better understanding of host-pathogen interactions and novel approaches can be developed to improve plant resistance against biotic stress (Chong and Xia, 2018; Chong et al., 2018; Misra et al., 2019).

Tools used for plant metabolomics

Plants are a rich and diverse source of metabolites and it has been suggested that a single plant contains 100,000 to 1 million metabolites. The study of plant metabolites is very important and singificant advances have been made in the last decade. A major attraction of this mode of study is identifying the regulatory importance of certain metabolites in the biochemical pathways of certain pathogens during host-pathogen interactions (Salem et al., 2020). In metabolomics, there is no single tool available to quantify all the metabolites present in the metabolome of a plant. As such, a set of analytical tools are employed in the analysis of metabolites. Different techniques are widely used to study the metabolome, such as mass spectrometry (MS), non-destructive nuclear magnetic resonance spectroscopy (NMR), high-performance thin-layer chromatography (HPTLC), capillary electrophoresis-mass spectrometry (CE-MS), gas chromatography-mass spectrometry (GC-MS), liquid chromatography-mass spectrometry (LC-MS), direct infusion mass spectrometry (DIMS), ultra-performance liquid chromatography (UPLC), high resolution mass spectrometry (HRMS), and Fourier transform ion cyclotron resonance mass spectrometry (FI-ICR-MS) (Razzaq et al., 2019). A number of factors, including accuracy, selectivity, speed, precision, and sensitivity, are all important when selecting the appropriate analytical tools for metabolomics.

Sample preparation plays a pivotal role in metabolomics because the credibility of the results depends on it. The collection, extraction, preparation, storage, and environmental conditions in which the sample is stored are all crucial. Even a slight change in temperature or pH may adversely affect the entire study. The Metabolomics Standards Initiative (MSI) should be followed for sample preparation, experimental design, and analysis (Sumner et al., 2007; Kim and Verpoorte, 2010; Cajka and Fiehn, 2016). Genome-wide association studies (GWASs) assisted by metabolomics techniques (mGWAS) and metabolic quantitative trait loci (mQTLs) are all powerful tools in detecting genetic variations linked with metabolic traits in plants (Gong et al., 2013). These are shown in [**31**].

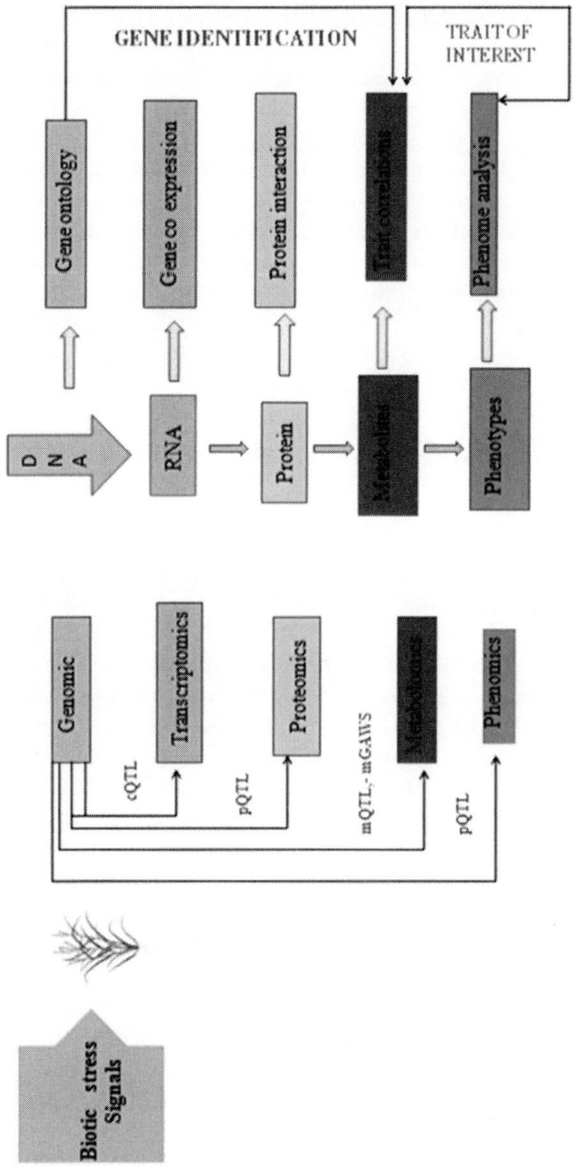

31. Genome-wide association studies (GWASs) assisted by metabolomics techniques (mGWAS) and metabolic quantitative trait loci (mQTLs) are powerful tools to detect genetic variations linked with metabolic traits in plants (Gong et al., 2013).

Mass spectrometry (MS)

Mass spectrometry (MS) quantifies the mass-to-charge proportion (m/z) of particles so as to recognize and evaluate atoms in simple and complex blends. It works with high throughput and large quantitative capacities to analyze the metabolites associated with various cell and organic pathways; it has likewise empowered putative biomarker disclosure in a few related examination zones. In-plant science, MS is a commonly used technique because of its easy connection with other separation methods, such as CE, GC, and LC. MS also has good accuracy, adaptability, and susceptibility. Direct infusion MS has low selectivity, which is considered a drawback of this analytical tool. When it is used in conjugation with GC, LC, and CE, selectivity is increased (Piasecka et al., 2019).

Gas chromatography-mass spectrometry (GC-MS)

GC-MS is widely employed to measure levels of volatile and semi-volatile organic compounds in biological samples. Alkylation, acylation, methoximation, trimethylsilylation, and silylation are commonly used methods of derivatization. The major advantage of GC-MS is its relative sensitivity and powerful reproducibility. It can measure hundreds of samples at a time. A GC-MS chromatogram contains many overlapping signals and peaks and it can be difficult to pick out the exact peak. The instrument is associated with several peak-selecting software packages with super-resolution functions (Obata and Fernie, 2012), which help make the technique cheaper and less time consuming. However, the use of GC-MS is restricted to thermostable and volatile compounds and it is difficult to analyze high molecular weight compounds and highly polar metabolites (Zaitsu and Hayashi, 2019).

Liquid chromatography-mass spectrometry (LC-MS)

Liquid chromatography combined with mass spectrometry is a well-accepted technique in plant metabolomics. Compared to GC-MS, it is analytically flexible in terms of polarity range and molecular weight with high resolution. The proportion of peaks in LC-MS depends on the type of sample and sample preparation methods. Accurate annotation of LC-MS data is pivotal to analysis (Dawid and Hille, 2018). Sophisticated and advanced tools are available for the correct annotation of data (Metaboanalyst). The greatest challenge in LC-MS analysis is the annotation of novel compounds (Shulaev, 2006). The database used for analysis is growing and its integration with other networks provides a powerful tool for analysis. Open spectral repositories with downloadable spectra and straightforward submission systems (e.g., MassBank and MoNa (MassBank of North America)) play a key role in facilitating accurate analysis. More refined software technologies and a broadening of the database may reduce the workload of researchers in coming years (Chaleckis et al., 2019).

Capillary electrophoresis-mass spectrometry (CE-MS)

CE-MS is a brand new technique with many advantages over LC-MS and GC-MS. It is highly sensitive and has high resolving power, low sample requirements, and short analysis time. It facilitates the simultaneous profiling of different classes of metabolites. As such, CE-MS is an attractive option for nontargeted metabolomic studies. The main drawback of CE-MS is its lower migration time and lack of reference libraries (Dawid and Hille, 2018).

High resolution mass spectrometry (HRMS)

Characterization of plant metabolites is rendered feasible with this stand-alone technique. It is highly accurate for mass resolution and mass measurement. HRMS possesses high mass resolving power and the ability to measure mass accurately in the ppm range. It generates very large experimental data sets that need to be evaluated subsequently using bioinformatics methods for data processing, data mining, and data curation. Nowadays, a large number of free data processing software packages are available, for instance XCMS. This software has multiple steps for data evaluation, including data extraction, alignment, and normalization. The interpretation of HRMS data is very complex because of its great size. Simplification of complex HRMS data is necessary for its accurate interpretation, which is made possible through the removal of interference peaks. Kendrick mass defect analysis is employed to process HRMS data. Van Krevelen diagrams provide another form of visual analysis of HRMS data. Mass defect filtering (MDF) is also employed for data processing to select ions of interest from a complex list. Some critical drawbacks faced by HRMS are the lack of its integration with automated data processing software packages and the lack of a mass spectral database (Rathahao-Paris et al., 2016).

Fourier transforms ion cyclotron resonance mass spectrometry (FI-ICR-MS)

This technique is used to study systems with large mass to charge ratios or having very complex isotope patterns. Here, the mass to charge ratio is measured by measuring the frequency at which the ion processes in a magnetic field. Compared to other mass spectrometry techniques, it gives high accuracy. Its high-resolution power contributes much to the accurate analysis of large metabolites in complex samples. It is widely used in metabolomics to study large molecules like proteins (Ghaste et al., 2016).

Nuclear magnetic resonance spectroscopy (NMR)

This is a powerful tool used to identify and quantitatively analyze plant metabolites either *in vivo* or in tissue extracts. It is comparatively less efficient than MS. To give coverage of the metabolome, complementary data from NMR is used along with MS. NMR is a versatile technique because it is non-destructive and spectra can be recorded from cell suspensions, tissues, and even whole plants, as well as from extracts and purified metabolites. The advantage of NMR as a metabolite profiling technique is that crude extracts can be used to obtain informative spectra; it is also possible to use inline separation techniques to simplify the analysis. This approach has had a major impact on the use of NMR for metabolic analysis. NMR is a potentially powerful strategy to reveal metabolome data on silent mutations in a phenotype through metabolite fingerprinting (Shulaev, 2006; Krishnan et al., 2005; Razzaq et al., 2019).

Other technologies

The challenge for researchers in metabolomics concerns the handling, processing, and analyzing of data, which requires specialized mathematical, statistical, and bioinformatical tools. Apart from analytical tools, several context-specific databases and tools have been developed (i.e. targeted on the integration of omics data from specific plant/animal models and medical/clinical studies). While most available databases are fairly general, there is also a range of databases with more specific focus. SOE species-specific databases are now available, such as the Plant Metabolic Network (PMN) (Schläpfer et al., 2017). Bioinformatics tools are crucial for further analysis of metabolomics data, including information management, raw analytical data processing, metabolomics standards and ontology, statistical analysis and data mining, data integration, and mathematical modeling of metabolic networks within the framework of systems biology. Among the reference resources providing metabolic and biological process information, the most widely used databases are KEGG, Reactome, and Metacyc (Ambrosino et al., 2020). In the following, some plant stress response databases are given: Arabidopsis thaliana Stress Responsive Gene Database (ASRGD); DroughtDB; PASmiR; PlantPReS; Plantstress.com; Plant Stress Gene Database; Plant Stress Protein Database (PSPDB); Rice SRTFDB; and the Stress Responsive Transcription Factor Database (STIFDB v.2). PSPDB is the plant stress protein database and is very useful to the scientific community researching biotic and abiotic stress responses (Anil Kumar et al., 2014). It contains details of 259 stress-related genes of 11 species, but only 33 genes from *Arabidopsis thaliana*. The Arabidopsis Stress Responsive Gene Database (ASRGDB) contains 636 records with 44 different types of stress response (Borkotoky et al., 2013). More and more stress-related genes, with related information and regular updates, will increase the usefulness of this database to the research community.

Application of metabolomics in agriculture

Plant metabolite content has importance for: plant development and differentiation; maturation of fruit; responses to environmental stress; and responses to biotic stress. Metabolomics can help us study the metabolite changes that occur during growth and maturation, which is useful for delaying fruit ripening and maturation. Stress response studies aid the development of resistant varieties. Host-pathogen interactions is another attractive area of metabolomics study and will aid in crop development and management. Metabolomic analysis is also recognized as a valuable tool in the risk assessment of genetically modified plants. Metabolomics has a significant role in the development of biomarkers for field applications and metabolomics technology is widely used in crop information measurement, disease diagnosis, etc., and contributes to increasing agricultural income (Ibarra-Estrada et al., 2016; Alseekh et al., 2018). The integration of the two has opened up a new direction for the development of the discipline of metabolomics. Translational metabolomics has contributed to the production of multifarious nutritional metabolites that have important roles in the human diet. Advances in metabolomics have shifted the technology from a focus on single metabolites to high throughput by generating milestones for a variety of metabolites, leading to the discovery of better metabolite networks (Kumar et al., 2017). As such, the application of metabolomics is improving the development of high performing crop genotypes to meet the challenges of 21st century agriculture.

References

Alseekh, S, Bermudez, L, de Haro, LA, Fernie, AR, and Carrari, F. 2018. "Crop metabolomics: from diagnostics to assisted breeding". *Metabolomics*, 14: 148.

Ambrosino, L, Colantuono, C, Diretto, G, Fiore, A, and Chiusano, ML. 2020. "Bioinformatics resources for plant abiotic stress responses: State of the art and opportunities in the fast evolving-omics era". *Plants*, 9: 591.

Anil Kumar, S, Hima Kumari, P, Sundararajan, VS, Suravajhala, P, Kanagasabai, R, and Kavi Kishor, PB. 2014. PSPDB: "Plant Stress Protein Database". *Plant Mol. Biol. Report*, 32: 940-942.

Atanasova-Penichon, V, Barreau, C, and Richard-Forget, F. 2016. "Antioxidant secondary metabolites in cereals: Potential involvement in resistance to Fusarium and mycotoxin accumulation". *Front. Microbiol.*, 7: 1-16.

Balmer, D, Flors, V, Glauser, G, and Mauch-Mani, B. 2013. "Metabolomics of cereals under biotic stress: Current knowledge and techniques". *Front. Plant Sci.*, 4: 82.

Bergelson, J, Kreitman, M, Stahl, EA, and Tian, D. 2001. "Evolutionary dynamics of plant R-genes". *Science*, 292: 2281-2285.

Borkotoky, S, Saravanan, V, Jaiswal, A, Das, B, Selvaraj, S, Murali, A et al. 2013. "The arabidopsis stress responsive gene database". *Int. J. Plant Genomics*, 2013: 1-3.

Boutrot, F, Zipfel, C. 2007. "Function, discovery, and exploitation of plant pattern recognition receptors for broad-spectrum disease resistance". *Annu. Rev. Phytopathol*, 55: 257-286.

Cajka, T, and Fiehn, O. 2016. "Toward Merging Untargeted and Targeted Methods in Mass Spectrometry-Based Metabolomics and Lipidomics". *Analytical Chemistry.* American chemical society: ACS publication.

Castro-moretti, FR, Gentzel, IN, Mackey, D, and Alonso, AP. 2020. "Metabolomics as an emerging tool for the study of plant-pathogen interactions". *Metabolites*, 10: 52.

Chaleckis, R, Meister, I, Zhang, P, and Wheelock, CE. 2019. "Challenges, progress and promises of metabolite annotation for LC-MS-based metabolomics". *Curr. Opin. Biotechnol.*, 55: 44-50.

Chen, D, Shao, Q, Yin, L, Younis, A, and Zheng, B. 2019. "Polyamine function in plants: Metabolism, regulation on development, and roles in abiotic stress responses". *Front. Plant Sci.*, 9: 1945.

Chong, J, Soufan, O, Li, C, Caraus, I, Li, S, Bourque, G. 2018. "MetaboAnalyst 4.0: Towards more transparent and integrative metabolomics analysis". *Nucleic Acids Res.*, 46: 486-494.

Chong, J, Yamamoto, M, and Xia, J. 2019. "MetaboAnalystR 2.0: from raw spectra to biological insights". *Metabolites*, 9: 57.

Chong, J, and Xia, J. 2018. "MetaboAnalystR: an R package for flexible and reproducible analysis of metabolomics data". *Bioinformatics*, 34: 4313-4314.

Dawid, C, and Hille, K. 2018. "Functional metabolomics-a useful tool to characterize stress-induced metabolome alterations opening new avenues towards tailoring food crop quality". *Agronomy*, 8: 138.

Ghaste, M, Mistrik, R, and Shulaev, V. 2016. "Applications of fourier transform ion cyclotron resonance (FT-ICR) and orbitrap based high resolution mass spectrometry in metabolomics and lipidomics". *Int. J. Mol. Sci.*, 17: 816.

Ghosson, H, Schwarzenberg, A, Jamois, F, and Yvin, JC. 2018. "Simultaneous untargeted and targeted metabolomics profiling of underivatized primary metabolites in sulfur-deficient barley by ultra-high performance liquid chromatography-quadrupole/time-of-flight mass spectrometry". *Plant Methods*, 14: 62.

Gong, L, Chen, W, Gao, Y, Liu, X, Zhang, H, Xu, C, Yu, S, Zhang, Q, Luo, J. 2013. "Genetic analysis of the metabolome exemplified using a rice population". *Proc. Natl Acad. Sci.*, 110: 20320-20325.

Hong, J, Yang, L, Zhang, D, and Shi, J. 2016. "Plant metabolomics: An indispensable system biology tool for plant science". *Int. J. Mol. Sci.*, 17: 767.

Hussain, SS, Ali, M, Ahmad, M, and Siddique, KHM. 2011. "Polyamines: Natural and engineered abiotic and biotic stress tolerance in plants". *Biotechnol. Adv.*, 29: 300-311.

Ibarra-Estrada, E, Soto-Hernández, RM, and Palma-Tenango, M. 2016. "Metabolomics as a Tool in Agriculture". In *Metabolomics-Fundamentals and Applications*, edited by Jeevan K. Prasin, 147-165. InTech publications.

Kasukabe, Y, He, L, Nada, K, Misawa, S, Ihara, I, and Tachibana, S. 2004. "Overexpression of Spermidine synthase enhances tolerance to multiple environmental stresses and up-regulates the expression of various stress-regulated genes in transgenic *Arabidopsis thaliana*". *Plant Cell Physio.*, 45: 712-722.

Kim, HK, and Verpoorte, R. 2010. "Sample preparation for plant metabolomics". *Phytochem. Anal.*, 21: 4-13.

Krishnan, P, Kruger, NJ., and Ratcliffe, RG. 2005. "Metabolite fingerprinting and profiling in plants using NMR". *Journal of Experimental Botany*, 56: 255-265.

Kumar, R, Bohra, A, Pandey, AK, Pandey, MK, and Kumar, A. 2017. "Metabolomics for plant improvement: Status and prospects". *Front. Plant Sci.*, 8: 1302.

Kushalappa, AC, and Gunnaiah, R. 2013. "Metabolo-proteomics to discover plant biotic stress resistance genes". *Trends Plant Sci.*, 18: 522-531.

Liu, Y, Guo, ., Ma, C, Zhang, D, Wang, C, Yang, Q. 2016. "Transcriptome analysis of maize resistance to *Fusarium graminearum*". *BMC Genomics*, 17: 477.

Misra, BB, Langefeld, C, Olivier, M, and Cox, LA. 2019. "Integrated omics: Tools, advances and future approaches". *J. Mol. Endocrinol.*, 62: 21-45.

Mittler, R. 2006. "Abiotic stress, the field environment and stress combination". *Trends Plant Sci.*, 11: 15-19.

Obata, T, and Fernie, AR. 2012. "The use of metabolomics to dissect plant responses to abiotic stresses". *Cell. Mol. Life Sci.*, 69: 3225-3243.

Piasecka, A, Kachlicki, P, and Stobiecki, M. 2019. "Analytical Methods for Detection of Plant Metabolomes Changes in Response to Biotic and Abiotic Stresses". *Int. J. Mol. Sci.*, 20.

Pinu, FR, Beale, DJ, Paten, AM, Kouremenos, K, Swarup, S, Schirra, HJ, et al. 2019. "Systems Biology and Multi-Omics Integration: Viewpoints from the Metabolomics Research Community". *Metabolites*, 9: 76.

Placinta, CM, D'Mello, JPF, and MacDonald, AMC. 1999. "A review of worldwide contamination of cereal grains and animal feed with *Fusarium mycotoxins*". *Anim. Feed Sci. Techno.*, 78: 21-37.

Rathahao-Paris, E, Alves, S, Junot, C, and Tabet, JC. 2016. "High resolution mass spectrometry for structural identification of metabolites in metabolomics". *Metabolomics*, 12: 1-15.

Razzaq, A, Sadia, B, Raza, A, Hameed, MK, and Saleem, F. 2019. "Metabolomics: A way forward for crop improvement". *Metabolites*, 9: 12.

Roessner, U, and Bowne, J. 2009. "What is metabolomics all about". *Beyond Darwin Futur. Mol. Biol.*, 46: 363-365.

Salem, M. A., De Souza, L. P., Serag, A., Fernie, A. R., Farag, M. A., Ezzat, S. M. 2020. "Metabolomics in the context of plant natural products research: From sample preparation to metabolite analysis". *Metabolites*, 10: 1-30.

Savary, S, Willocquet, L, Pethybridge, SJ, Esker, P, McRoberts, N, and Nelson, A. 2019. "The global burden of pathogens and pests on major food crops". *Nat. Ecol. Evol.*, 3: 430-439.

Schlaeppi, K, and Mauch, F. 2010. "Indolic secondary metabolites protect Arabidopsis from the oomycete pathogen *Phytophthora brassicae*". *Plant Signal. Behav.*, 5: 1099-1101.

Schläpfer, P, Zhang, P, Wang, C, Kim, T, Banf, M, Chae, L. 2017. "Genome-wide prediction of metabolic enzymes, pathways, and gene clusters in plants". *Plant Physiol.*, 173: 2041-2059.

Scholz, M, Gatzek, S, Sterling, A, Fiehn, O, and Selbig, J. 2004. "Metabolite fingerprinting: Detecting biological features by independent component analysis". *Bioinformatics*, 20: 2447-2454.

Shulaev, V. 2006. "Metabolomics technology and bioinformatics". *Brief. Bioinfor.*, 7: 128-139.

Shulaev, V., Cortes, D., Miller, G., and Mittler, R. 2008. "Metabolomics for plant stress response". *Physiol. Plant*, 132: 199-208.

Solomon, PS, and Oliver, RP. 2001. "The nitrogen content of the tomato leaf apoplast increases during infection by *Cladosporium fulvum*". *Planta*, 213: 241-249.

Stahl, EA, and Bishop, JG. 2000. "Plant-pathogen arms races at the molecular level". *Curr. Opin. Plant Biol.*, 3: 299-304.

Stahl, EA, Dwyer, G, Mauricio, R, Kreitman, M, and Bergelson, J. 1999. "Dynamics of disease resistance polymorphism at the Rpm1 locus of Arabidopsis". *Nature*, 400: 667-671.

Sumner, LW, Amberg, A, Barrett, D, Beale, MH, Beger, R, Daykin, CA. 2007. "Proposed minimum reporting standards for chemical analysis: Chemical Analysis Working Group (CAWG) Metabolomics Standards Initiative (MSI)". *Metabolomics*, 3: 211-221.

Zaitsu, K, and Hayashi, Y. 2019. "GC/MS/MS based targeted metabolomics for pathophysiological analysis of animal models". *Med. Mass Spectrom.*, 3: 11-18.

Zeiss, DR, Mhlongo, MI, Tugizimana, F, Steenkamp, PA, and Dubery, IA. 2019. "Metabolomic Profiling of the Host Response of Tomato (*Solanum lycopersicum*) Following Infection by *Ralstonia solanacearu*m". *Int. J. Mol. Sci.*, 20: 3945.

Zhou, H, Lin, J, Johnson, A, Morgan, RL, Zhong, W, and Ma, W. 2011. "Pseudomonas syringae type III effector HopZ1 targets a host enzyme to suppress isoflavone biosynthesis and promote infection in soybean". *Cell Host Microbe*, 9: 177-186.

Zhu, L, Zhou, Y, Li, X, Zhao, J, Guo, N, and Xing, H. 2018. "Metabolomics Analysis of Soybean Hypocotyls in Response to Phytophthora sojae Infection". *Front. Plant Sc.*, 9: 1530.

CHAPTER NINE

INFLUENCE OF MAGNETIC NANOPARTICLES ON PLANT GROWTH AND STRESS ADAPTATION IN NATURAL ENVIRONMENTS

ABHIJITH P, BHAVYA G AND GEETHA N

Magnetic fields are a universal phenomenon—just as hydrogen is the most abundant element in the universe, magnetic fields are almost omnipresent. Magnetic fields are present from the centre of the universe up to its known boundaries. Magnetism is a source of life in its manifestation as morphogenic fields within cells or taking the form of gravity—it makes life possible on our planet. The interactions of these magnetic fields with plants has been studied since the 1900s with J. C. Bose being a pioneer in this area of study. Bose studied plant responses to electrical fields. The magnetic field, along with the electrical field, has a net influence on plants and they behave according to the net impact of external stimuli. These initial observations paved the way for researchers to further investigate matters concerning the interaction of magnetic fields role with plants to create both physical and molecular changes. Tiny circular regions of magnetism can be swiftly increased with the intervention of precise measuring of the magnetic properties of nanoparticles, similar to the expansion of bubbles in a just-opened bottle of champagne. The diverse relationships of magnetic fields and living cells have been studied in a broader and deeper way in terms of the magnetic behavior of nanoparticles in areas including: seed germination; development; seedling treatment; enzymatic changes; morphological characteristics; yield; fertilizer uptake; early growth of plants; and biotic and abiotic stress tolerance. These studies all give a well-rounded picture of how both natural and manmade magnetic nanoparticles, acting as tiny actuators, can directly influence how plants respond and behave on exposure, and give us a glimpse into a future of technological possibilities in electro-nanomagnetic agricultural systems for the improvement of crop production and protection, and providing a host of opportunities and applications previously unimaginable.

Introduction

All interactions in nature are governed by four basic physical forces. Of these four, electromagnetic plays a vital role in how the universe is shaped. That being said, it is clear that a force of such importance interacts significiantly with life as we know it.

Every living organism is in constant interaction with the electromagnetic fields that occur in and around it during every moment of its life, be it in terms of metabolism (enzyme kinetics), embryo development (morphogenic fields), or wound healing. These complex interactions are very subtle and studies were initiated early in the 20th century. The famous Indian scientist, Dr. Jagadish Chandra Bose, was a pioneer in studying the responses of plants to different electromagnetic stimuli (Tandon, 2019; Bose, 1918). He developed an instrument known as a crescograph to study the subtle movements in plants upon stimulation by electromagnetic fields. In the modern age, the study of magnetism has lead us from studies of basic science to the pinnacle of scientific knowledge and we now know that plants are not just affected by electromagnetic radiation, but are moulded by it. The germination of seeds, early growth and development, flowering and fruit setting, yields, responses to both biotic and abiotic stress, and even senescence are all interconnected with the electromagnetic fields to which plants are subjected (Bose, 1918).

Magnetic fields are powerful in themselves, but their study experienced a turning point with the identification of magnetic nanoparticles. These tiny particles of <100 nm in size play a significant part in the physico-chemical interactions that happen inside plants. Nanoparticles play a major role in disease resistance and the modulation of stress responses in plants, and it has been suggested that, in the future, nanopesticides will revolutionize the agriculture sector. Magnetic nanoparticles are tiny in size (giving a very high overall surface area and better interactions at the cellular level) and have magnetic properties, as well as resonance interactions. It is possible to target any part of a living organism using magnetic nanoparticles to give targeted delivery of drugs or chemical molecules; to generate electromagnetic resonance at a target site inducing a specific genetic expression; or to control how a protein behaves within the magnetic field. These intricate interactions show us how much of this world is being shaped by electromagnetic forces and make us wonder "is this the invisible hand of god".

The history of magnetism and magnetic resonance in plant science

Magnetic fields are everywhere, constantly shaping and reshaping the world around us. Birds use the earth's magnetic field to find their way in the open sky; fish use it in the ocean to find their way; and before the advent of the global positioning system (GPS), humans used the magnetic compass to navigate, which relies on earth's magnetic field. Magnetic fields play a vital role in every aspect of our lives and all life on our planet.

Magnetic fields interact with animals and also have tremendous effects on the growth of plants. Sir Jagadish Chandra Bose, the famous Indian physicist, science fiction writer, and biologist, studied the responses of plants to electric stimuli. He invented the crescograph, a device to measure growth and movement in plants in response to electrical stimuli. His experiments paved the way to better understanding of plant responses to electromagnetic stimuli, which has turned into an exciting area of research.

In 1960, Krylov and Tarakonova reported an auxin-like effect on seed germination in seeds exposed to magnetic fields. Later, in 1963 the same auxin-like effect was seen in the ripening of tomato fruits by Boe and Salunkhe. Some experiments have even suggested that the root growth response is linked to magnetotropism or geo-magnetotropism, not just to heliotropism. Even though these early works have given us a lot of insights, the very fact that earth's geomagnetic fields affect everything around us made it hard to conduct many experiments with plants and magnetic fields in the early days. Today, as more accurate devices have been devised, research in the area of field interactions in plants have gained momentum once again, especially in the new century.

A number of recent studies (2004 to 2016) have suggested that the growth patterns of forests and many specific species of plants are directly linked to the magnetic field strength on the earth's surface. Occhipinti et al. (2014) note subtle changes in the earth's surface magnetic fields ranging from 30 uT in South American and South African regions up to 70 uT in northern Canada, the south of Australia, and parts of Siberia. Qamili et al. (2013) described the main source of the magnetic fields on earth as internally originating fields caused by the dynamo-like effect of the earth's core and molten metallic layers and the secondary source found in the ionosphere and magnetosphere surrounding earth's atmosphere.

While the effects of magnetic fields are obvious, how these interactions happen are of interest to the scientific community. While there have been many explanations proposed by major scientific teams around the world, the only plausible explanation most scientists seem to accept is that of a resonance-based interaction. Resonance is the change in state of a magnetic dipole or absorption or emission of electromagnetic radiation in response to an applied electromagnetic field. In simpler terms, electromagnetic resonance can induce a current in a coil when it is placed in an oscillating magnetic field without any physical contact—this is how dynamos, generators and transformers work.

Biomagnetic fields

Magnetobiology covers the study of magnetic fields and their interactions and effects on living organisms. It explores the phenomenon of biomagnetism—magnetic fields produced by living organisms. Its use in navigation and perception is described as magnetoception (Maffei, 2014). To better explain, we can take the example of the electroencephalograph (EEG). This device measures the electrical pulses in the brain, which are of extremely low intensity, but provide a clear picture of how the brain is functioning. Similarly, a magnetoencephalography (MEG) is used to study the minute magnetic fields produced by the human brain (Singh, 2014). More than a 160 laboratories around the world are studying the minute magnetic fields produced by human bodies and other living organisms. The tricky part is that, to study such minute fields, one needs a magnetically shielded room so that only fields of interest can be studied without external interference.

The effects of extremely low frequency (ELF) electromagnetic fields (EMF) have been studied thoroughly in the last decade (Maffei, 2014). These studies generally point us towards an understanding of how the ultra-low intensity magnetic fields contribute to the growth and development of living cells. The term "morphogenetic magnetic fields" is now being used to describe the low intensity, low frequency electromagnetic waves that contribute to the development and morphological change of organisms, ranging from single cells to blue whales. Many studies on embryos suggest that the entire body plan is derived from the electromagnetic fields of single cells and their interactions with the fields of neighboring cells. Collectively, all this work helps us to explore the intricate patterns of nature at work and how ultra low biomagnetic fields and their complex integration weave the fabric of reality, yielding the biological world we see around us.

Magnetism and magnetic materials

Magnetism is the net result of the movement of charges. Whenever a charged particle moves in space, it creates a magnetic field, and this makes it a fundamental phenomenon in nature. Each atom has a cluster of electrons encircling the nucleus and each of these electrons carry a charge—their net movement constitutes a magnetic field. As such, we can say that every single atom is a tiny magnet of its own making, which further highlights how magnetic fields are of the utmost importance to life. Every living thing is built up of these tiny atomic magnets and the ways they interact with external magnetic fields give us life as we know it. This is the story at the microscopic level; at the macro level, we can see how these magnetic fields protect us. For example, the magnetosphere, which wraps around the entire planet, shields us from solar winds and an artillery barrage of sub-atomic particles that could be deadly if it were not kept in check.

Magnetic materials are classified according to the behavior of a material with magnetism. They are primarily categorized according to the alignment of their magnetic moments under an applied magnetic field: **diamagnetic**; **paramagnetic**; **ferromagnetic**; **ferrimagnetic**; and **antiferromagnetic** (De silva et al., 2000) [**32**].

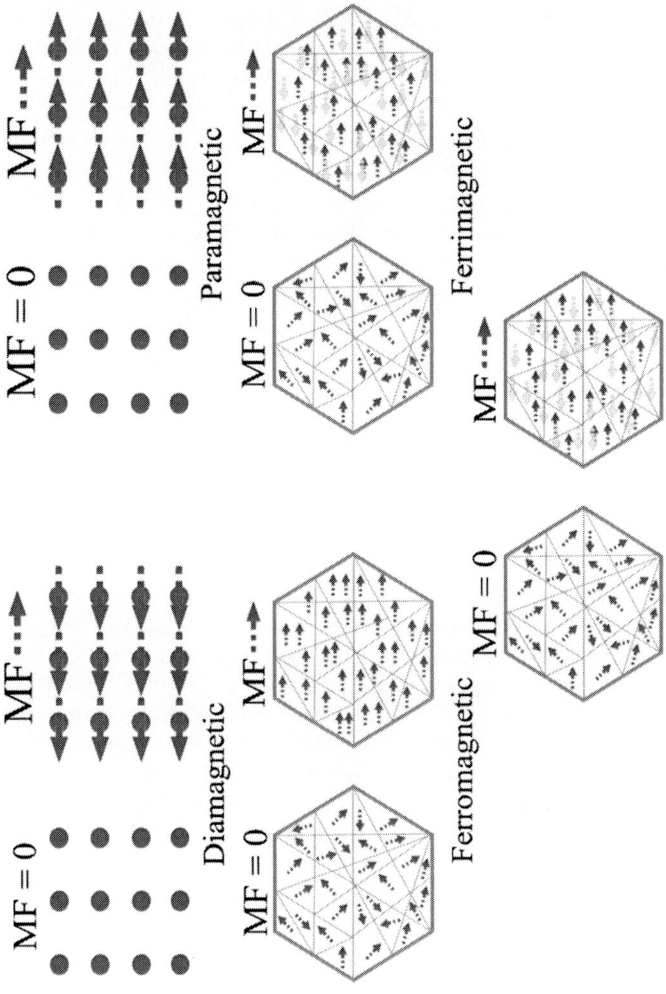

32. Alignment of magnetic moments in different materials under an applied external magnetic field.

Diamagnetic materials are repelled by magnetic fields. The applied magnetic field induces an opposite magnetic field in the material, which causes a repulsive force. Atoms in a diamagnetic material have no net magnetic dipole moment. All materials are diamagnetic in nature, but the force is very weak in general. **Paramagnetic** materials have an unpaired electron in their partially filled orbitals, which constitutes a net magnetic moment. Due to this, when an external magnetic field is applied there is a partial positive alignment of the atomic magnetic moments creating an attractive force, which creates positive magnetization and positive susceptibility. However, when the applied field is removed, the net magnetic force disappears. **Ferromagnetic** materials possess a permanent magnetic moment due to the atoms having a lot of unpaid electrons in their orbitals. These materials (such as iron, magnetite (Fe_3O_4), etc.) are very powerful magnets when an external magnetic field is applied to them. The magnetism in ferromagnetic material is the result of the alignment patterns of their constituent atoms, which behave as elementary electromagnets. In general, ferromagnetic materials form the strongest magnets possible.

Ferromagnetic materials possess permanent magnetism. Their atomic magnetic fields spontaneously align to a specific configuration thus creating a permanent magnetic moment for the material. Some atomic magnetic fields align in parallel and some align anti-parallel within the same material, which causes magnetic dilution. As such, these materials do not form magnets as powerful as ferromagnetic materials. The magnetic alignment of these materials is disrupted beyond a specific temperature, known as the Curie point, where the magnet will lose its magnetic properties—this is a property of every material in this category. When the temperature is brought back down below the Curie point, the ferromagnetic properties are revived.

Anti-ferromagnetic materials have a magnetic alignment that is more or less temperature dependant. At a specific temperature, known as the Néel temperature, anti-ferromagnetic materials exhibit magnetic behavior. The magnetic moments, in pairs, are aligned in parallel or anti-parallel directions in the normal state and are mostly canceled out by each other. As the temperature nears the Néel temperature, some atoms break out of the order and start to align with the external magnetic field causing a specific magnetic moment. This action is maximized at the Néel temperature and disappears completely over the specified temperature. These properties are taken into consideration and materials are chosen to produce the right magnetic nanoparticles. It is crucial to understand the behaviors of these materials under the specific conditions of the applied external magnetic field. The making of this choice is very important, as different materials have different magnetic behaviors depending on the temperature and behavior of atomic magnets (Rikken et al., 2014; Da silva et al., 2000).

Magnetic nanoparticles

A particle with a size of less than 100 nanometers [**33**] is usually considered a nanoparticle. Magnetic nanoparticles have similar magnetic properties like ferromagnetism. Iron oxide nanoparticles (Fe_3O_4) are one of the most important groups of nanoparticles. These are ferromagnetic nanoparticles that exhibit exceptional superparamagnetism at the nanoscale. These particles can be coated with active compounds, like phospholipids or fatty acids, and then be used to incite specific stress tolerance responses in plants. Once these particles are given to a plant, we can use an external magnetic field to guide them to a specific area and target a disease or pathogen. These particles are exceptionally powerful systems to deliver and induce specific plant responses as needed.

33: Scanning electron micrograph of metallic nanoparticles at 15000x magnification (Source: Abhijith et al., 2019).

Over the years, there have been many efforts to make iron oxide particles more efficient. Such methods include surface functionalization, physico-chemical stabilization, and optimal sizing etc. For a particle to be most effective in the biological system, it is best to have a size of 50 to 100 nm, which also helps improve their magnetic properties. Stability is a major concern with iron oxide nanoparticles. Uncoated particles are easily susceptible to degradation by oxidation or agglomeration, which reduces their efficiency over time. The best strategy is to use a coating to ensure the particles do not degrade while in storage. Coating with multiple layers of organic or inorganic compounds ensures better overall stability and the coating can also serve as a functional surface layer for use as a catalyst or biomarker. In theory, a size range of 50 to 160 nm and a high magnetization value is what matters most for a magnetic nanoparticle. When coated with the right compounds, such a particle will have optimal stability and biocompatibility.

Biological and synthetic methods can both be used produce these particles. High quality iron oxide magnetic nanoparticles can be produced by thermal decomposition, co-precipitation, micelle formation, and hydrothermal and laser pyrolysis techniques. The method is chosen according to need, as these synthetic methods can be used to control the size, shape, and dispersion characteristics of the resulting magnetic nanoparticles. For practical applications, more than 90 % of MNPs are produced through synthetic methods [**34**].

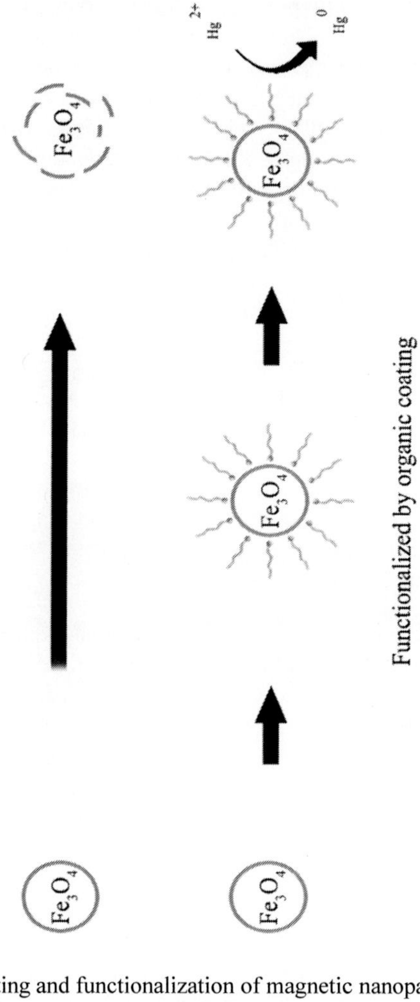

34: Coating and functionalization of magnetic nanoparticles.

Plant growth in magnetic fields

Plants growing naturally in the wilderness are affected by geomagnetic fields (GMF). According to Maus et al. (2006), the earth's magnetic field, with a value of up to approximately 5×10^{-6} Tesla (T), is produced by a molten core of iron rotating inside the earth's core. The magnetic field ranges from 25 to 65 μT according to the inclination and declination and is the naturally occurring component of the environment in which the plant grows. Even though it is natural, this magnetic field can cause some abiotic stress of the plant (Wang et al., 2006). Despite being a natural and widespread phenomenon that has been studied since the1930s, few studies were done in this area until the past decade. It is known that factors such as the magnetic field strength, duration, frequency, and orientation can cause direct changes in plant metabolism and growth. Galland and Pazur (2005) categorized magnetic fields into four major types: weak static homogeneous magnetic fields (0-100 μT, including GMF); strong homogeneous magnetic fields (milli-Tesla to μTesla); strong inhomogeneous magnetic fields; and ELF magnetic fields of low to moderate (several hundred μT) magnetic flux densities [**35**].

Magnetic fields have been tested as a physical pre-treatment in seed germination experiments, which have shown that they help improve seed germination, seedling emergence, and increased initial growth. The major properties that affect seedlings include the frequency, magnetic flux density, exposure time, absolute exposure dose, and the polarity of the applied field. Changing these parameters according to the genotype of the plant gives optimal results.

To study the effects of super weak (near zero) magnetic fields, the absence of the geomagnetic field is required. This requires either shielding or compensating for the GMF. Neamţu and Morariu (2005) conducted a number of studies in a controlled environment with selective radiation of super weak magnetic fields and showed an increase in the germination of rye seeds of up to 133 %. The study also showed that a zero-magnetic field condition (ZMF) has negative effects on root growth, which was corrected when a super weak magnetic field was reintroduced to the plant. Fischer et al. (2004) noted that exposure to a weak vertical magnetic field of 16 2/3 Hz and sinusoidal 20 μT for 12 days increased the fresh weight of shoots and roots in potted marigold plants. Celestino et al. (2000) used a Helmholtz coil system to irradiate cork oak (*Quercus suber* L.) acorns with EMF at 50 Hz, 15 μT for 13 weeks and found that the number and shoot and root weight increased significantly. A study by Cakmak et al. (2010) found that wheat and common bean seeds exposed to a 7 mT magnetic field for 7 days showed improved seedling growth and germination, even under saline conditions and increased osmotic pressure. Similarly, exposure to low strength magnetic fields was found to induce seed growth under conditions of cold stress and drought; increased seed germination was even seen under low humidity conditions.

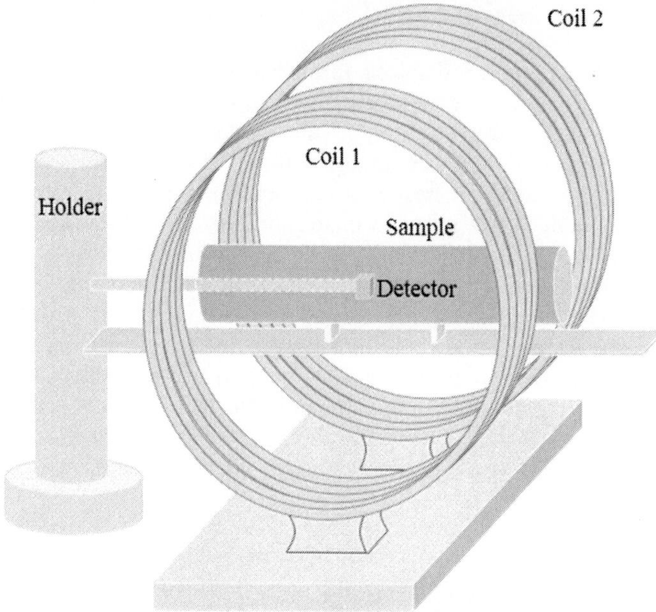

35. Helmholtz coil system (He and Hong, 2018).

These results suggest a potential method for further increasing the efficiency of such magnetic fields by supplementing the plant with magnetic nanoparticles and then applying similar magnetic fields. Ultra-low magnetic fields show better efficiency when the plant is supplied with iron oxide nanoparticles. Both root and shoot growth was shown to increase when supplemented with iron oxide NPs (Wang et al., 2011; Liu et al., 2016). The same treatment was also shown to increase the production of wet and dry biomass weight in plants. Ren et al. (2011) noted that direct application to soil of iron oxide NPs gave increased root growth of up to 50.3 %. The effects of iron oxide NPs are not limited to growth enhancement; they also increase germination rate, disease resistance, stress tolerance, and yield. Many similar experiments have ben conducted that clearly highlight the direct implications of magnetic fields on plant growth, stress tolerance, and yield. All of this points towards the possibility of a future system in agriculture in which the way we produce crops is revolutionized by the use of magnetic stimuli and magnetic nanoparticles, offering an eco-friendly means of crop improvement.

Stress adaptation of plants in magnetic fields

Stress in both its biotic and abiotic forms has negative impacts on plants in terms of growth, disease resistance, and yield. Mitigating the negative impacts of stress and making plants more tolerant has long been a very important aspect of research in agricultural science. Salinity, drought, waterlogging, irradiation, heavy metals, and extreme temperatures, both high and low, all induce stress in plants and lead to the overproduction of reactive oxygen species (ROS), such as hydroxyl radicals, hydrogen peroxide, hydroperoxy radicals, superoxide radicals, and singlet oxygen, which can be very harmful to plants (Khan et al., 2017). At higher concentrations, these free radicals cause nucleic acid, protein, and lipid membrane degradation, in turn causing cytotoxicity. Răcuciu et al. (2008) explain that magnetic fluid exposure can induce stress in plants at specific frequencies and can also stimulate the biosynthesis of nucleic acids to remarkable levels, as plant adaptations to magnetic fields see different effects.

The use of magnetic iron oxide nanoparticles, with their large surface area and small size, presents a solution to these issues. Iron oxide NPs can help reduce ROS activity by scavenging them. Their large surface area creates more space for absorbing and removing ROS from the system as they are produced. Research has also pointed to the ability of MNPs to reduce oxidative and osmotic stress and increase photosynthesis (Haghighi and Pessarakli, 2013; Qi et al., 2013; Siddiqui et al., 2014). These capabilities are enhanced, as plants grow naturally in a low intensity magnetic field, by the likes of the earth's geomagnetic field (GMF), which enhances the potency of the particles in question. The effect of iron oxide NPs is the net result of not only their particle size, but also the greater surface area they provide. They can maximize their effects due to their potent ferromagnetic properties, which, together with the earth's GMF, give rise to a more stable magnetic field application in plants.

Researchers found that, when used as a foliar spray, nano-sized maghemite, an iron oxide MNP, increased drought tolerance in *Helianthus* annuus and *Carthamustinc torius* L. (Bameri et al., 2013; Zareii et al., 2014; Martínez-Fernández et al., 2015; Aghdam et al., 2016). They also found that yield was also increased in some occasions. A foliar spray of Fe_2O_3 MNPs helps combat salinity by reducing the accumulation of Na^+ and Cl^- ions in plants; it can also help upregulate the nutritional uptake of plants, with a net effect of combating salinity and reducing ROS production (Soliman et al., 2015; Khan et al., 2017). A chilling treatment of surface functionalized

NPs improved biochemical and physiological activities in plants under cold stress (Haghighi et al., 2014; Kohan and Geisler, 2015). All these studies point towards the possibility of providing a complete array of magnetic field and magnetic nanoparticle-based technologies that can revolutionize agriculture.

Magnetic NPs in plant growth and stress management

MNPs are crucial to plant growth and management under stress conditions. The size of MNPs is a very important factor for their amalgamation and efficient transfer into the biological system (Al-Almri et al., 2020). The application of high concentrations of magnetic Fe_3O_4 nanoparticles to *Ipomoea aquatica* plants in a hydroponic system increased their water utilization capacity and secondary metabolite production, but decreased the plants' growth rate and assimilation of carbon (Yusufirashim et al., 2019). On the contrary, the accumulation of Fe_3O_4 NPs in plant cells activated the plant immune response mechanism by upregulating the salicyclic acid (SA) responsive genes PR1 and PR2. This activated enhanced resistance towards TMV infection by increased antioxidative enzymes and the abscisic acid (ABA) hormone in *Nicotiana benthamiana*. The Fe_3O_4 NPs also acted as a plant growth promoter and micronutrient supplement, as well as an elicitor against tobacco mosaic virus (TMV), a plant pathogen (Chai et al., 2020). The aqueous root application of Fe_3O_4 nanoparticles in *Triticum aestivum* L. displayed a super paramagnetic signal without hampering the growth rate and chlorophyll content. Localization was restricted to the root epidermal region and nowhere else. The Fe_3O_4 NPs induced the activation of the antioxidant system in the root and aerial parts, thus acting as preventive oxidative molecules (Iannone et al., 2016). The increased expression of miR159c has a well-known role in the resistance mechanisms of plants against fungal pathogens. Fe_3O_4 NPs effect the expression of the miR159c gene. One study highlighted its increased expression in *Medicago falcate* L. The activity of miR159c on Fe_3O_4 NPs depends on the plant species (Kokina et al., 2020) [**36**].

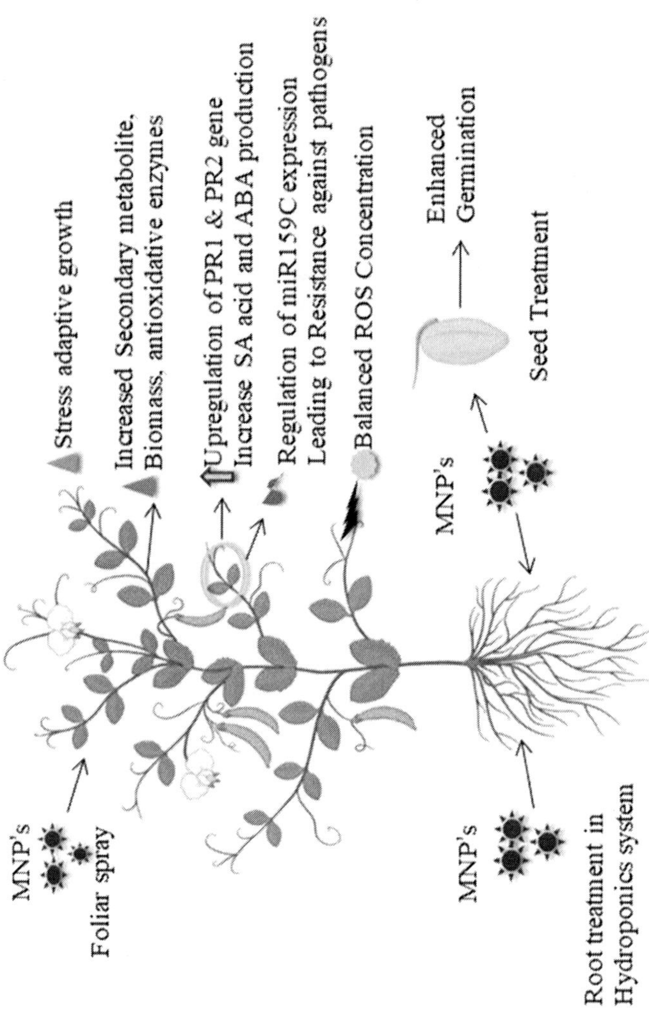

36: The role of MNPs in plant growth and stress management (Basic plant structure created with BioRender.com).

Future prospects for MNPs and resonance in combating the global food crisis

The current system of agriculture is clearly only borderline effective. It has a number of disadvantages that have made themselves very clear over the years as we have learned of the carcinogenicity of most of the fertilizers and pesticides used. It is high time we find a solution that is more stable and sustainable for both nature and mankind. Polluted food will not give us a better future, but only more diseases. Furthermore, many areas in the world are now seeing increasingly stressful conditions from global warming, changes in the water table, increased salinity, and higher disease prevalence—we have been trying to combat these factors with chemicals that date from the early 1970s. The time has come to take the next step in organic agronomic methods and make the change from chemical-based agriculture to safer magnetic field-based practices. The advent and use of magnetic nanoparticles have shown us how they can be used to combat biotic and abiotic stresses. With the advantages of small size and large surface area, MNPs offer the best solution.

The direct use of surface functionalized magnetic NPs has been shown to elicit some good responses in tandem with low intensity GMF. The practicalities of applying an external field at a magnitude of a few micro teslas to an entire field using a large solenoid system, like a functional magnetic fence system through which a specific frequency magnetic field can be applied, is challenging. MNPs are highly ferromagnetic and can easily pick up this external field, becoming more magnetic and inciting better and stronger effects in plants. MNPs can be stabilized by surface coatings, making it sufficient to apply the particles once every two months or so through an aerial spray and then use coils encircling the crop field to radiate the magnetic field. They could be used to treat specific infections by surface coating the right compounds and using a large area magnetic field to guide the particles to a specific location after aerial spraying. This would give us more control over how and where the particles act and ensure minimal dosage; the applied magnetic field would also boost the plants' defense systems. Mahajan and Pandey (2014) have described how magneto-priming of seeds can increase resistance to infection. All this research has clearly shown that the future of agriculture lies in the hands of nanoparticles and magnetic treatments, which will be able to help plant growth in almost all stressful conditions and help reduce the harm caused by the use of pesticide and chemicals.

All these aspects point to a better mode of agriculture where we can grow food in almost any conditions at lower costs and with better safety. This will help us some way to resolving the world's food crisis, as stress and scarcity of nutrient resources is a large problem that negatively impacts global food production, along with the prevalence of diseases. The use of optimized MNPs and related technologies can help stabilize food production to better scale up and meet the requirements of our growing population.

Conclusions

Magnetic nanoparticles can answer many of the questions that have eluded scientists in the field of agriculture and can bring about changes that will answer some of the problems currently faced. Once we address the "how to" of handling the stress conditions properly, then growing food in any part of the world under all conditions becomes a much easier task. Added to this, an understanding of how magnetic fields interact with plants during growth opens us up to new possibilities of terraforming other planetary bodies, like the moon and mars with radically different magnetic dynamics. The understanding we gain from these studies will pave the way to setting up cultivation systems on other planets as we explore and colonize them in the future. These studies will surely be of the utmost use in learning how we can create infrastructure that can help us cultivate plants in another gravity field on a different planet or in deep space while traveling for lightyears. Technology has always created the space for new inventions that can bring about radical changes in every field known to mankind; from the wheel to the space age we have continued to evolve scientifically.

Nanobiotechnology has brought forth disruptive and never before seen changes in the field of biology over the past 20 years. The merger of scientific disciplines to produce more interdisciplinary prespectives and research has created more knowledge, which will propel us to a state that we can call 'Civilization 4.0'. This may not be the final point, as the future is forever changing, but it is the next small step for man and a greater step for mankind into the future.

References

Abhijith, P, G Bhavya, P, Ashika, N, Geetha. 2019. Characterization of Silver Nanoparticles From the endangered medicinal plant *Salacia Chinensis* and analysis of its Bioactive potentials. *Life Science Edge*, Vol. 5, No. 1: 43-50.

Aghdam, Mohammad Taieb Baiazidi, Hamid Mohammadi, and Mansour Ghorbanpour. 2016. "Effects of nanoparticulate anatase titanium dioxide on physiological and biochemical performance of *Linum usitatissimum* (Linaceae) under well-watered and drought stress conditions". *Brazilian Journal of Botany*, 39(1): 139-146.

Al-Amri N, Huseyin Tombuloglu, Yassine Slimani, Sultan Akhtar, Mohammad Barghouthi, Munirah Almessiere, Thamer Alshammari et al. 2020. "Size effect of iron (III) oxide nanomaterials on the growth, and their uptake and translocation in common wheat (*Triticum aestivum* L.)". *Ecotoxicology and Environmental Safety*, 194: 110377.

Bameri M, Rooholla Abdolshahi, Ghasem Mohammadi-Nejad, Khatoon Yousefi, and Sayed Masoud Tabatabaie. 2012. "Effect of different microelement treatment on wheat (*Triticum aestivum*) growth and yield". *International Research Journal of Applied and Basic Sciences* 3(1): 219-223.

Boe, AA, and DK Salunkhe.1963. "Effects of magnetic fields on tomato ripening". *Nature*, 199, No. 4888: 91-92.

Bose, JC. 1908. "Mechanical and electrical response in plants". Dublin: Read before the British Association.

Cai, Lin, Liuti Cai, Huanyu Jia, Changyun Liu, Daibin Wang, and Xianchao Sun. 2020. "Foliar exposure of Fe_3O_4 nanoparticles on *Nicotiana benthamiana*: Evidence for nanoparticles uptake, plant growth promoter and defense response elicitor against plant virus". *Journal of Hazardous Materials* 122415.

Cakmak, Turgay, Rahmi Dumlupinar, and Serkan Erdal. 2010. "Acceleration of germination and early growth of wheat and bean seedlings grown under various magnetic field and osmotic conditions". *Bioelectromagnetics: Journal of the Bioelectromagnetics Society*, 31, 2: 120-129.

Celestino, C, ML Picazo, and M Toribio. 2000. "Influence of chronic exposure to an electromagnetic field on germination and early growth of *Quercus* suber seeds: preliminary study". *Electro and Magnetobiology*, 19, No. 1: 115-120.

Da Silva, XA, IS Oliveira, and AP Guimarães. 2000. "Ideal Types of Magnetic Materials." *Revista Brasileira de Ensino de F sica*, 22, 3: 306-311.

Davar, Zareii Farinaz, Arash Roozbahani, and Amir Hosnamidi. 2014. "Evaluation the effect of water stress and foliar application of Fe nanoparticles on yield, yield components and oil percentage of safflower (*Carthamus tinctorious* L.). *Int. J. Adv. Biol. Biomed. Res.*, 2: 1150-1159.

Fischer, G, M Tausz, M Köck, and Dieter Grill. 2004. "Effects of weak 162\over3 Hz magnetic fields on growth parameters of young sunflower and wheat seedlings". *Bioelectromagnetics: Journal of the Bioelectromagnetics Society*, 25, No. 8: 638-641.

Galland, Paul, and Alexander Pazur. 2005. "Magnetoreception in plants". *Journal of Plant Research*, 118, No. 6:371-389.

Haghighi, Maryam, and Mohammad Pessarakli. 2013. "Influence of silicon and nano-silicon on salinity tolerance of cherry tomatoes (*Solanum lycopersicum* L.) at early growth stage". *Scientia Horticulturae*, 161: 111-117.

Haghighi, Maryam, Reza Abolghasemi, and Jaime A. Teixeira da Silva. 2014. "Low and high temperature stress affect the growth characteristics of tomato in hydroponic culture with Se and nano-Se amendment". *Scientia Horticulturae*, 178: 231-240.

He, Jie, and Hong Ye. 2018. "Measurement and Analysis of Low Frequency Magnetic Shielding Performance of Open-Cell Nickel Foam". *MS&E* 394, No. 3: 032122.

Iannone, María Florencia, María Daniela Groppa, María Elisa de Sousa, Marcela Beatriz Fernández van Raap, and María Patricia Benavides. 2016. "Impact of magnetite iron oxide nanoparticles on wheat (Triticumaestivum L.) development: evaluation of oxidative damage". *Environmental and Experimental Botany*, 131: 77-88.

Khan, M, Nasir, M, Mobin, Zahid Khorshid Abbas, Khalid A Al-Mutairi, and Zahid H Siddiqui. 2017. "Role of nanomaterials in plants under challenging environments". *Plant Physiology and Biochemistry*, 110: 194-209.

Kohan-Baghkheirati, Eisa, and Jane Geisler-Lee. 2015. "Gene expression, protein function and pathways of *Arabidopsis thaliana* responding to silver nanoparticles in comparison to silver ions, cold, salt, drought, and heat". *Nanomaterials*, 5, No. 2: 436-467.

Kokina, Inese, Ilona Plaksenkova, Marija Jermaļonoka, and Anastasija Petrova. 2020. "Impact of iron oxide nanoparticles on yellow medick

(*Medicago falcata* L.) plants". *Journal of Plant Interactions*, 15, No. 1: 1-7.

Krylov, AV, and Tarakonova, GA. 1960. "Magnetotropism of plants and its nature". *Fiziologlya Rastienji*, 7: 917-919.

Liu, Ruiqiang, Huiying Zhang, and Rattan Lal. 2016. "Effects of stabilized nanoparticles of copper, zinc, manganese, and iron oxides in low concentrations on lettuce (*Lactuca sativa*) seed germination: nanotoxicants or nanonutrients?". *Water, Air, & Soil Pollution* 227, No. 1: 42.

Maffei, Massimo E. 2014. "Magnetic field effects on plant growth, development, and evolution". *Frontiers in Plant Science*, 5: 445.

Martínez-Fernández, Domingo, Didac Barroso, and Michael Komárek. 2016. "Root water transport of *Helianthusannuus* L. under ironoxide nanoparticle exposure". *Environmental Science and Pollution Research*, 23, No. 2: 1732-1741.

Maus, S, Martin Rother, Claudia Stolle, W, Mai, S, Choi, Hermann Lühr, D, Cooke, and C, Roth. 2006. "Third generation of the Potsdam Magnetic Model of the Earth (POMME)". *Geochemistry, Geophysics, Geosystems* 7, No. 7: Q07008.

Neamţu S, and VV Morariu. 2005. "Plant growth in experimental space flight magnetic field conditions". *Romanian Journal of Biophysics*, 15: 41-46.

Occhipinti, Andrea, Angelo De Santis, and Massimo E. Maffei.2014. "Magnetoreception: an unavoidable step for plant evolution?". *Trends in Plant Science*, 19, No. 1: 1-4.

Qamili, E, A De Santis, A Isac, M Mandea, B Duka, and A Simonyan. 2013. "Geomagnetic jerks as chaotic fluctuations of the Earth's magnetic field". *Geochemistry, Geophysics, Geosystems*, 14, No. 4: 839-850.

Qi, Mingfang, Yufeng Liu, and Tianlai Li. 2013. "Nano-TiO 2 improve the photosynthesis of tomato leaves under mild heat stress". *Biological Trace Element Research*, 156, No. 1-3: 323-328.

Răcuciu, Mihaela, Simona Miclăuş, and Dorina-Emilia Creangă. 2009. "The response of plant tissues to magnetic fluid and electromagnetic exposure". *Romanian J. Biophys.*, 19: 73-82.

Ren, Hong-Xuan, Ling Liu, Chong Liu, Shi-Ying He, Jin Huang, Jun-Li Li, Yu Zhang, Xing-Jiu Huang, and Ning Gu. 2011. "Physiological investigation of magnetic iron oxide nanoparticles towards Chinese mung bean". *Journal of Biomedical Nanotechnology*, 7, No. 5: 677-684.

Rikken, Roger, SM Roeland, JM Nolte, Jan C Maan, Jan CM van Hest, Daniela A Wilson, and Peter CM Christianen. 2014. "Manipulation of micro-and nanostructure motion with magnetic fields". *Soft Matter*, 10, No. 9: 1295-1308.

Siddiqui, Manzer H, Mohamed H Al-Whaibi, Mohammad Faisal, and Abdulaziz A Al Sahli. 2014. "Nano-silicon dioxide mitigates the adverse effects of salt stress on Cucurbita pepo L.". *Environmental Toxicology and Chemistry* 33, No. 11: 2429-2437.

Singh, Sanjay P. 2014. "Magnetoencephalography: basic principles". *Annals of Indian Academy of Neurology*, 17, No. Suppl. 1: S107-S112.

Soliman, Amira Sh, Souad A El-feky, and Essam Darwish. 2015. "Alleviation of salt stress on *Moringaperegrina* using foliar application of nanofertilizers". *Journal of Horticulture and Forestry*, 7, No. 2: 36-47.

Tandon, Prakash Narain. 2019. "Jagdish Chandra Bose & plant neurobiology". *The Indian Journal of Medical Research*, 149, No. 5: 593.

Wang, BC, Jing Zhou, YC, Wang, LC, Zhu, and JA, Teixeira da Silva. 2006. "Physical stress and plant growth." Floriculture, ornamental and plant biotechnology: advances and topical issues 3: 68-85,1st edn. Global Science Books Ltd, Isleworth, 68-85

Wang, Huanhua, Xiaoming Kou, Zhiguo Pei, John Q. Xiao, Xiaoquan Shan, and Baoshan Xing. 2011. "Physiological effects of magnetite (Fe_3O_4) nanoparticles on perennial ryegrass (*Lolium perenne* L.) and pumpkin (*Cucurbita mixta*) plants". *Nanotoxicology* 5, No. 1: 30-42.

Yusufirashim, RM, Ibrahim, MH, Che Azurahanim, Che Abdullah, and Izad AA. 2019. "Growth, Photosynthesis and Quality of Water Spinach (*Ipomoea aquatica*) as Influenced by Magnetic Nanoparticles (MNP) Application". *Annual Research & Review in Biology*, 31(6): 1-15.

CHAPTER TEN

MOBILE RADIATION RESPONSES IN PLANTS

P. DUHAN

The use of mobile phones is increasing every day. Mobile phones have become one of the fastest growing technologies in the world. Mobile phones emanate electromagnetic radiation into the environment, having tremendous effects on human and plant health. Radiation is not always hazardous. The lethality of radiation depends on how long something is exposed to it and what kind of radiation it is. A number of studies have been made in various countries on the impact of mobile phone radiation on plants. Preliminary research has revealed that electromagnetic radiation emitted from mobile phones inhibits root growth, physiology, morphology, and biochemistry through the induction of oxidative damage. The inhibitory effect of radiation was found to be greater on root growth than on shoot growth. It has also been observed that electromagnetic radiation from mobile phones may stunt the growth of agricultural crops and plants. This radiation can affect morphological and biochemical changes in plants. It can also affect fruit, such as guava, grapes, apple, and coconut, as well as flowering plants, such as jasmine and chilli plants, etc. Several strategies can be adopted to check electromagnetic pollution in the natural environment and prevent environmental harm caused by electromagnetic radiation in plants and other agricultural crops, such as that emitted by mobile phones. In this chapter, the changes in plants and crops when they come into contact with mobile radiation are discussed.

Introduction

Radiation is a part of everyday life and is emitted from both natural sources, like the sun, and artificial sources, such as broadcast towers, remote controls, and electrical and electronic equipment like cellphones and TVs. Radiation can be classified into two types: ionizing and non-ionizing. Ionizing radiation has a short wavelength/high frequency and high energy; it is usually harmless. Non-ionizing radiation has a relatively long wavelength, low frequency, and low energy. Due to its higher energy, ionizing radiation (IR) can break chemical bonds, produce ions, and can even damage and destroy the nucleus of an atom and directly affect an organism's DNA resulting in dangerous mutations.

Radiation damage depends on the time of exposure and type of radiation. According to the International Telecommunications Union (ITU), worldwide there are about 7.377 billion mobile telephone subscriptions (ITU, 2017). Nowadays, much of the population of the world uses cellphones for communication without considering the side effects. Mobile phones and towers emit signals of very low frequency (824-1780 MHz), which is called microwave radiation [**37**]. A cellphone has a specific absorption rate (SAR) (Kumar, 2010). Cellphones and towers emit signals via radiowaves of mobile frequency (MF), which is a form of electromagnetic radiation (EMR). This radiation has a harmful effect on both animals and plants and is carcinogenic for human beings (Smith et al., 2000). Manmade EMR has become so ubiquitous that it is now increasingly being recognized as a form of unnoticed and dangerous environment pollutant (Mahajan et al., 2012). When plants come into contact with mobile radiation, they undergo various types of morphological and biochemical change, such as altered gene responses, chromosomal aberrations, a reduction in the reproduction rate, DNA damage, decreased growth, direct burn damage to exposed tissue, and ultimately mortality of the plant [**38**]. Direct contact with radiation is not required to affect plant life—the mere presence of radiation is often enough (Miller, 2015). The radiation emitted by mobile phones may affect plants in both a positive and negative manner. The consequences of mobile phone radiation on plants are the subject of recent interest due to the enormous increase in mobile phone usage throughout the world.

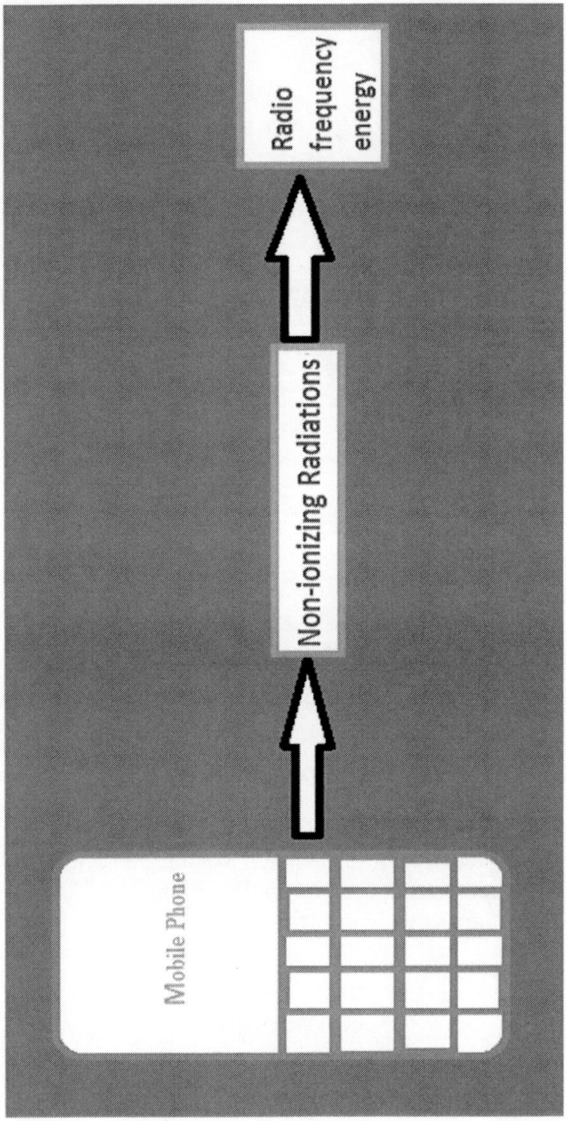

37. Radiation emitted by cellphones.

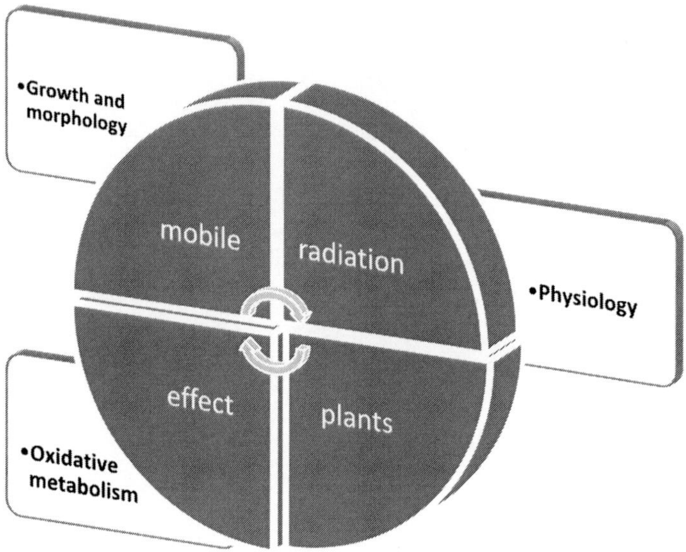

38. Mobile phone radiation-effects on plants.

Effects on germination, growth, morphology, and physiology

Growth is the basic phenomenon of living organisms. In plants, growth is defined as an increase in mass and volume. Previous studies have suggested that radiation emitted by mobile phones can clog seeds, affect germination, and impact early growth, ultimately affecting plant morphology and physiology.

Germination

Mobile phone radiation affects germination in both positive and negative ways. A number of researchers have found a negative impact of radiation on seed germination and early growth in *Lens culinary*, *Vigna radiate*, *Triticum*, *Brassica*, *Phaseolus aureus*, and *Vigna aconitifolia* (Akbalet al., 2012; Afzal and Mansoor, 2012; Singh et al., 2011; Martin, 2009; Bellad and Kubihal, 2015). Akbal et al. (2012) found noticeable effects of mobile radiation on dormant seeds, rather than in a state of germination. On exposure to mobile phone radiation (1800 MHz), dormant *Lens culinaris Medik* seeds saw a reduction in seedling root growth (60 %) and mitotic index (12 %). Seed root growth decreased possibly as a result of oxidation. Likewise, a reduction in growth in *Vigna* and *Triticum* (21 and 50 % respectively) was observed under cellphone radiation of 900 MHz for 4 hours (Afzal and Mansoor, 2012). In addition, cell phone EMFr was also shown to inhibit the germination and early growth of *Brassica* (Singh et al., 2011). Sharma et al. (2010) subjected *Vigna radiata* to cellphone radiation (900 MHz) and found it impaired early growth by inducing biochemical changes. 50 % inhibition of germination was also reported. A few studies have found a positive effect of mobile phone radiation on germination. Sharma and Parihar (2014) observed a considerable increase in germination percentage, seedling length, and plant size in seedlings of *Pisum sativum* and *Trigonella foenumgraecum* exposed to mobile phone radiation.

Growth and morphology

Similarly, both positive and negative effects have been observed in roots and shoots. A significant reduction was found in seedlings emerging from cellphone-exposed seeds. The inhibitory effect of electromagnetic radiation was greater on root growth than on shoot growth (Martin, 2009). In *Brassica*, root growth was also inhibited by reactive oxygen species (ROS) generating oxidative stress in spite of the increased activity of antioxidant enzymes (Singh et al., 2011). Root and stem elongation were severely affected in *Phaseolus aureus* and *Vigna radiate* after exposure to cellphone radiation (Sharma et al., 2009, 2010).

Halgamuge (2016) assessed 29 different species of plants to see the effect of mobile non-thermal, weak, radiofrequency electromagnetic fields (RF-EMF). His study showed physiological and/or morphological changes in plants. Additionally, analysis of the results of these studies demonstrates that maize, roselle, pea, fenugreek, duckweed, tomato, onion, and mung bean plants seem to be very sensitive to RF-EMFs. Other morphological changes, like variations in nodule size and number, were observed in fenugreek and pea seedlings after exposure to mobile phone radiation. An increase in both the number of nodules and size in both pea and fenugreek seeds was seen over an extended time period (Sharma and Parihar, 2014). Another study conducted by Gupta on chilli plants found that those exposed to mobile radiation (from the seed stage to the growth stage) attained greater height and longer leaves. However, the leaves were wrinkled and looked dull compared to normal plants. Sharma et al. (2010) found dry weight significantly reduced by mobile radiation to 43 % in *Vigna radiate* after exposure for 4 hours. A similar result was observed by Bellad and Kubihal (2015) in moth bean (*Vigna Aconitifolia*) and wheat (*Triticum aestivum*). This reduction was associated with increased membrane damage and antioxidant enzyme activity.

Physiology

A few reports are available on the affect of mobile radiation on plant physiology. A highly significant reduction in relative water content (RWC) was observed by Bellad and Kubihal (2015) in moth bean (*Vigna Aconitifolia*) and wheat (*Triticum aestivum*). However, an alteration in the conformation of biomolecules, such as proteins, nucleic acids, and membrane lipids, was observed due to high energy microwave radiation, which can break the chemical bonds due to low quantum energy and thermal effect (Caldwell et al., 1995; Barnes and Cardoso-Vilhena, 1996). Ha et al. (1997) and Havaux (1998) discover how modifications in the biomolecular tertiary structure can alter the rate of physiological processes.

Alterations to oxidative metabolism

In addition to other changes, mobile phone radiation induced production of ROS has been observed in plants and this was thought to be the primary cause of DNA strand breaks and cell death (Oktem at al., 2005; Oral et al., 2006; Sokolovic et al., 2008). ROS include superoxide radicals (O_2^-), hydroxyl radicals (OH^-), and H_2O_2, which cause oxidative stress in plants through the peroxidation of lipids, proteins, and nucleic acid, leading to membrane disintegration followed by cellular leakage and cell death. Lipid oxidation causes the production of MDA and is measured in terms of MDA content. Afzal and Mansoor 2012 estimated MDA content in mung bean (*Vignaradiata*) and wheat (*Triticumaestivum*) seedlings under controlled conditions and in the presence of mobile phone radiation. They found a significant increase in MDA content in stressed wheat and mung bean plants. A significant reduction in protein content was observed in both wheat and mung bean under mobile radiation stress conditions with a higher decrease in wheat (Afzal and Manssor, 2012). Zare et al. (2015) also observed the effect of electromagnetic waves with high radio frequency radiation on the antioxidant activities of *Zea mays* and found an increase in MDA content in the leaves of plants under stress conditions.

Sharma et al. (2010) exposed 7-day old seedlings of *Vigna radiata* to cell phone EMF for 0.5, 1, 2, and 4 hours (a 5th set was maintained under control conditions). A significant reduction in root protein content was observed in all sets except the control and an increasing reduction in protein content was found with increased duration of exposure. There was a maximum reduction of 58 % after 4 hours of exposure to cellphone EMF. The generation of free radicals (ROS) has been considered a primary cause for this reduction of protein content (Monselise et al., 2003). Singh and Prakash (2011) also noticed an increase in MDA content in plants under cell phone EMF stress conditions. Tkalec et al. (2007) also observed similar results in *Lemna minor*. Chandel et al. (2017) exposed onion roots to continuous wave homogenous EMFr for 1, 2, and 4 hours for a single day and observed ROS generation in stressed plants. In addition, MDA content was also found to increase in stressed onion roots. Sharma et al. (2013) also noticed mobile phone induced generation of H_2O_2 and lipid peroxidation in *Vigna radiata* roots. The level of oxidative stress was sufficiently high that root growth was inhibited even with increased activity of antioxidative enzymes. Antioxidative enzymes and non enzymatic substances play a vital role in oxidative metabolism, as these scavenge ROS, prevent cellular damage, and help plants cope with stressful conditions. There have been several reports of increasing antioxidative enzyme activity in response to cellphone radiation. Afzal and Mansoor (2012) observed increased antioxidative enzymes, such as guaiacol peroxidase (GPX), ascorbate peroxidase (APX), and catalase (CAT). Wheat and mung bean seedlings were exposed to GSM mobile phone (900 MHz) band frequency with modulated voice and low frequency signals. Sharma et al. (2010) observed increased activity of proteases, alpha-amylases, beta-amylases, polyphenol oxidases, and peroxidases in EMF-exposed radicles and found them helpful for the survival of plants under stress conditions, indicating their role in protecting against EMF induced stress. Tkalec et al. (2005) studied the effect of electromagnetic fields on *Lemna minor* and found an increase in POX in plants in an electromagnetic field of 900 MHz for 2 hours. Tkalec et al. (2007) also observed increased catalase activity in *Lemna minor* when exposed to non-thermal EMF of 400 and 900 MHz. An increase in the activity of antioxidant enzymes, superoxide dismutases (SOD) and catalases (CAT), was observed in onion roots when exposed to 2100 MHz cellphone EMF (Chandel et al., 2017). A significant upregulation in the activity of scavenging enzymes was noted in mung bean roots when exposed to mobile phone radiation (Sharma et al., 2013). Based on the above studies, it is clear that mobile phone radiation has an impact on oxidative metabolism by affecting ROS and antioxidative enzyme content.

Conclusions and future perspectives

Mobile phone use is constantly increasing. Farmers throughout the world use phones without understanding their effects on crops. Mobile towers are also located in farmers' field and these produce RF, a type of non–ionizing electromagnetic radiation. This mobile radiation shows a significant impact on plants. A lot of work has been carried out on understanding the impact of EMFr on plants at the cellular, molecular, genetic, and whole plant level. However, the extent of the effect depends on the exposure of the plant to radiation. As such, more advanced research needs to be done to develop strategies that ameliorate its harmful effects.

References

Afzal, M, Mansoor, S. 2012. "Effect of mobile phone radiations on morphological and biochemical parameters of mung bean (*Vignaradiata*) and wheat (*Triticum aestivum*) seedlings". *Asian Journal of Agricultural Sciences*, 4(2): 149-152.

Akbal, A, Kiran, Y, Sahin, A, Turgut-Balik, D, Balik, HH. 2012. "Effects of electromagnetic waves emitted by mobile phones on germination, root growth, and root tip cell mitotic division of *Lens culinaris* Medik". *Polish Journal of Environmental Studies*, 21(1): 23-29.

Barnes, JD, Cardoso-Vilhena, J. 1996. "Interactions between electromagnetic radiation and the plant cuticle". In Kerstiens G (ed.) *Plant Cuticles. An Integrated Functional Approach.* Bios Scientific Publishers: Oxford, 157-174.

Bellad, A, Kubihal, CV. 2015. "Effect of Cell Phone Radiations on Morphological and Biochemical Parameters of Moth Bean (*Vignaaconitifolia*) and Wheat (*Triticumaestivum*): GSM vs CDMA". *The Indian Journal of Nutrition and Dietetics*, 52(4): 390-396.

Caldwell, MM, Teramura, AH, Tevini, M, Bornman, JF, Björn, LO, Kulandaivelu, G. 1995. "Effects of increased solar ultraviolet radiation on terrestrial plants". *AMBIO: A Journal of the Human Environment*, 24: 166-173.

Chandel, S, Kaur, S, Singh, HP, Batish, DR, Kohli, RK. 2017. "Exposure to 2100 MHz electromagnetic field radiations induces reactive oxygen species generation *llium cepa* roots". *Journal of Microscopy and Ultrastructure*, 5(4): 225-229.

Ha, MA, Apperley, DC, Jarvis, MC. 1997. "Molecular rigidity in dry and hydrated onion cell walls". *Plant Physiology*, 115: 593-598.

Halgamuge, MN. 2017. "Review: Weak radiofrequency radiation exposure from mobile phone radiation on plants". *Electromagnetic Biology and Medicine*, 36(2): 213-235.

Havaux, M. 1998. "Carotenoids as membrane stabilizers in chloroplasts". *Trends Plant Science*, 3: 147-151.

Kumar, G. 2010. *Cell Tower Radiation*. Electrical Engineering Department IIT Bombay, Powai, Mumbai, 3-32.

Mahajan Ankur et al. 2012. "Human Health and Electromagnetic Radiations". *International Journal of Engineering and Innovative Technology*, 1(6): 95-97.

Martin, M. 2009. "Mobile radiation stunts crop growth". *India Today*.

Monselise, E, Parola, BHA, Kost, D. 2003. "Low-frequency electromagnetic fields induce a stress effect upon higher plants, as evident by the universal stresssignal, alanine". *Biochemical and Biophysical Research Communications*, 302: 427-434.

Oktem, F, Ozguner, F, Mollaoglu, H, Koyu, A, Uz, E. 2005. "Oxidative damage in the kidney induced by 900MHz-emitted mobile phone: protection by Melatonin". *Arch. Res.*, 4: 350-535.

Oral, B, Mehmet, G, Fehmi, O, Nermin, K, Tamer, M, Selcuk, C and Gokhan, C. 2006. "Endometrial apoptosis induced by a 900-MHz mobile phone: Preventive effects of vitamins E and C". *Advances in Therapy*, 6: 957-973.

Rammal, M, Jebai, F, Rammal, H, Joumaa, WH. 2014. "Effects of long term exposure to RF/MW radiations on the expression of mRNA of stress proteins in *Lycopersicon esculentum*". *WSEAS Transactions on Biology and Biomedicine*, 11: 10–14.

Reed Miller. 2015. "Effects of Radiation on Plants". Submitted as coursework for PHD 241, Stanford University.

Sharma, VP, Singh, HP, Kohli, RK, Batish, DR. 2009. "Mobile phone radiation inhibits *Vigna radiata* (mung bean) root growth by inducing oxidative stress". *Science of the Total Environment*, 407(21): 5543-5547.

Sharma, S, Parihar, L. 2014. "Effect of mobile phone radiation on nodule formation in the leguminous plants". *Current World Environment*, 9(1): 145-155.

Sharma, VP, Singh, HP, Batish, DR, Kohli, RK. 2010. "Cell phone radiations affect early growth of vigna radiate (Mung Bean) through biochemical alterations". *Zeitschrift für Naturforschung C*, 65(1-2): 66-72.

Sharma, VP, Singh, G, Kohli, RK. 2009. "Effect of mobile phone EMF on biochemical changes in emerging seedlings of *Phaseolus aureus* Roxb". *The Ecoscan*, 3(3-4): 211-214.

Singh, Pal H, Sharma, Prakash, V. 2011. "Electromagnetic radiations from cell phone inhibit plant root growth through induction of oxidative damage". *Chemicals and environmental Health Issues Epidemiology*, 22: 249-252.

Smith, C, Collins, D. 2000. *3G Wireless Networks*. 136.

Sokolovic, D, Djindjic, B, Nikolic, J, Bjclakovic, G, Pavlovic, D, Kocic, G, Kristic, D, Cvetkovic, T, Pavlovic, V. 2008. "Melatonin reduces oxidative stress induced by chronic exposure of microwave radiation from mobile phones in rat brain". *The Journal of Radiation Research*, 49 (6): 579-586.

Tkalec, M, Malaric K, Pevalek-Kozlina, B. 2005. "Influence of 400, 900, and 1900 MHz electromagnetic fields on *Lemna minor* growth and peroxidase activity". *Bioelectromagnetics*, 26(3): 185-193.

Tkalec, M, Malaric, K, Pevalek-Kozlina, B. 2007. "Exposure to radiofrequency radiation induces oxidative stress in duckweed *Lemna minor* L.". *Science* of *the Total Environment*, 388(1-3): 78-89.

Zare, H, Mohsenzadeh, S, Moradshahi, A. 2015. "Electromagnetic waves from GSM a mobile phone simulator and abiotic stress in *Zea mays* L.". *Journal of Nutrition & Food Sciences*, S11: 3.

CHAPTER ELEVEN

PROGRAMMED CELL DEATH AND ABIOTIC STRESS RESPONSES IN PLANTS

RK UPADHYAY AND L DARNEI

Just as for animals, death is an integral part of the lifecycle of plants. Cell death is essential for the growth and development of eukaryotes, maintaining tissue and organ homeostasis in concert with cell proliferation, growth, and differentiation. Programmed cell death is defined as any form of cell death involving a single or a series of molecular and cellular orderly processes mediated by intracellular death programmes, regardless of the (external) trigger or the hallmarks they exhibit (Dangl et al., 2000). In plants, PCD occurs during many developmental processes and under several abiotic stress conditions. For instance, PCD has been reported under hypoxia stress (Drew et al., 2000) and extreme temperature (Vacca et al., 2004); PCD-related responses to abiotic stresses in animals and plants display both standard and distinct features. Additional studies aimed at determining cellular proteins fragmented by phytaspase, as well as the elucidation of the mechanism underlying retrograde phytaspase trafficking in the course of PCD, can provide valuable insights into the molecular mechanisms that underpin plant responses to stress. Some plants have also developed the ability to adapt to anaerobic stress, which is paradoxically aimed at the selective death of root and shoot cells (Drew et al., 2000). This mechanism relies on the pivotal role of PCD, starting with anaerobically sprouted seedlings that emerge from water and come into contact with atmospheric air, forming the plant's root system. This results in the formation of continuous air cavities (aerenchyma) that readily enable oxygen transport from aboveground aerated organs to the roots and the survival of the whole plant under unfavourable conditions. Programmed cell death, as a type of organized cell death, also functions in response to several other abiotic stresses, such as salinity, cold, heat, and oxidative stress. These stress induced responses are of considerable interest, not only from a fundamental viewpoint, but also in the context of agriculture, ecology, and environmental protection. As such, elucidation of the mechanisms and the specific enzyme systems involved in these processes may lay the foundation for future development of biotechnological methods enabling the generation of plants that are more tolerant to these stress conditions.

Overview

In plants, programmed cell death is an active cell process that occurs during different developmental stages and in response to environmental cues. It occurs as a dynamic process and displays predictable patterns throughout development and in response to pathogenic infection. PCD plays a vital role in plants, though aspects of its function and mechanism differ. In plants, PCD is essential for development and survival. For example, xylem vessels, which are composed of cells that compose the elements of the water and nutrient transport system, die at maturity, but remain because of PCD, leaving behind strong walls. Plants also employ PCD in response to pathogens. Various ROS species and cellular redox changes can trigger different signaling cascades leading to PCD. In plants, in addition to mitochondria, chloroplasts are also necessary for ROS suppliers and may generate intermediate signals involved in PCD processes. In addition to the participation of mitochondria, the active involvement of chloroplast-derived signals is also important during plant PCD. A strong interplay between ROS and other signaling molecules (phytohormones) is found in plant PCD (Overmyer et al., 2005). Moreover, ROS dependent PCD is associated with increased levels of both salicylic acid and ethylene, as well as the excessive generation of an SA-degrading enzyme (catechol) in an ethylene mutant, which has been found to reduce PCD. This firmly positions both ethylene and SA in a positive feedback cycle that promotes ROS dependent cell death (Danon et al., 2005).

Occurrence

PCD occurs dramatically in plant tissues. It plays a systematic and vital role and its presence has been observed at many stages of plant reproduction. The most significant PCD event is found in the vascular systems of plants, where water and nutrient conducting tubes are formed. Havd and Durzan, in the reproduction of Norway spruce during somatic embryogenesis, observed how nuclei slated to die displayed DNA strand breaks with 3'OH ends and apoptotic bodies. Two tassel seed genes in maize required for cell death during sex determination are closed down in this species. Sex determination involves the selective killing of female reproductive structures for the survival of male floral arrangements that may develop in the tassel.

Recent work by Cheung and colleagues has suggested that PCD also occurs in the transduction tissue through which pollen tubes grow. The process is selective since the tissues surrounding the transduction tissue in the pistil remain intact. Cell activity appears to be correlated with pollen tube growth, since incompatible pollen does not elicit cell death. PCD also occurs during reproductive development in plants. The cells of the tapetum, the cell layer that surrounds the developing pollen grains in the anthers, undergo a precise programme of breakdown that results in the release of their content, providing nutrients for the plant. The best-studied examples of ROS-derived PCD are those following the typical biphasic oxidative burst during both HR and the response to ozone stress. The ensuing cell death is characterized by discrete cellular lesions that are preceded by the appearance of several hallmarks of PCD, such as chromatin condensation, DNA laddering, and cytochrome c release (Lam, 2004).

Additional evidence for ROS dependent PCD in plants has been provided by transgenic plants deficient in catalase (Cat1AS), which is the major H_2O_2 scavenging enzyme. In plants, in addition to mitochondria, chloroplasts are also necessary for supplying ROS and can generate intermediate signals involved in PCD. For example, during cryptogenic induced PCD, H_2O_2 dependent activation of lipoxygenases targets chloroplastic polyunsaturated fatty acids, thus releasing oxylipins that are sufficient for inducing PCD (Montillet et al., 2005). Therefore, redox dependent plant PCD involves a range of signaling molecules, but this interactive regulatory network is only now starting to be elucidated [**39**].

Functions

PCD performs an essential function in plants, although its role remains to be determined in a number of cases. PCD, in its most dramatic occurrence in plants, results in the development of the vascular system. Secondly, if vascular bundles are severely wounded such that parenchyma cells are induced to redifferentiate into xylem, then PCD results in tracheary xylem elements functioning as long conducting tubes to transport water from the roots to the rest of the plant and providing mechanical support. Third, many tissues experience damage caused by pathogen invasion, resulting in

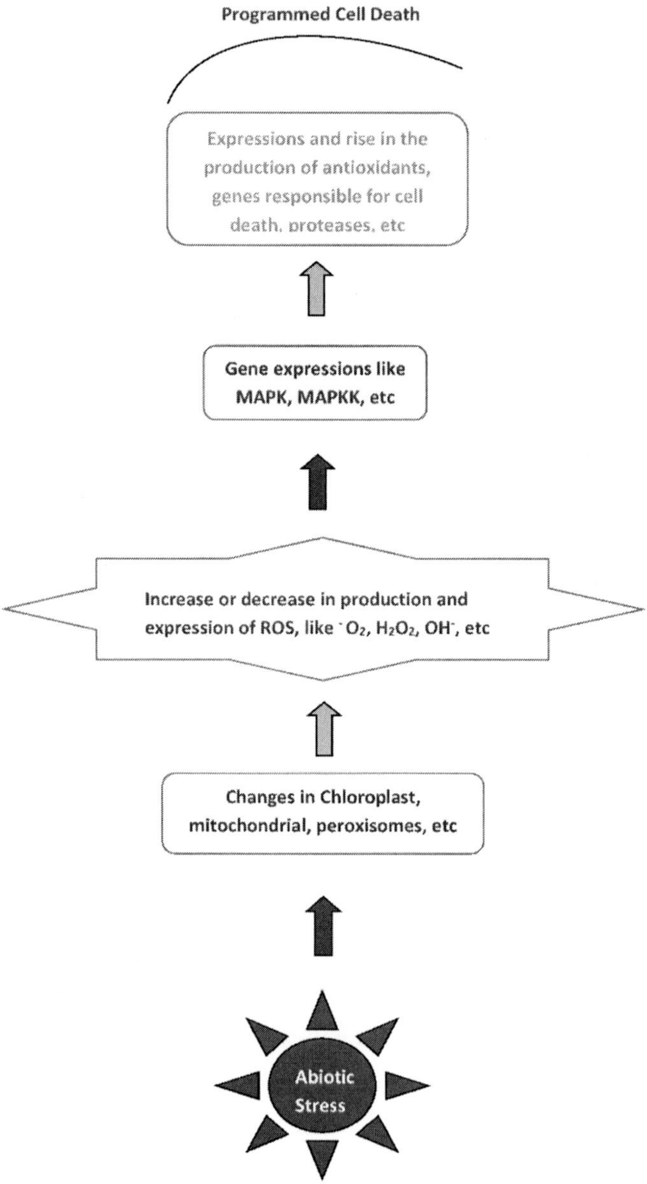

39: Abiotic stress responses to programmed cell death in plants.

the disruption of normal function. In such cases, PCD protects the tissues from pathogen attack. Fourth, though its role during pollen tube elongation has not been firmly established, PCD results in the provision of nutrients to growing cells. Fifth, pollen tubes require many nutritional elements during their growth. In some instances, the other elements needed to support normal function.

As such, PCD allows the physical accommodation of growing pollen tubes. Sixth, a dramatic hallmark of the resistance response of plants is the induction of a localized cell death response at the site of infection. A hypersensitive reaction is a widespread phenomenon that is responsible for the activation and establishment of plant immunity to certain diseases. Hence, the HR is an excellent example of programmed cell death, leading to rapid, localized cell death at infection sites. It contributes to the limitation of growth and the spread of the invading pathogen.

The functions and necessity of PCD for plant development and defense determine the underlying regulatory mechanism and process by which PCD is executed. Without its occurrence, in some instances, plants would be unable to perform their normal functions and would struggle to survive [**40**]. It leads to the revival of different parts of the plant body and regenerates them accordingly. Cell death occurs predictably at specific sites and time points throughout the life history of flowering plants. For example, the cells of the suspensor, the embryonic organs that attach the embryo proper to the maternal tissues and supply it with nutrients, die before the embryo matures. In particular examples, such as that of *Monstera* sp., PCD plays a vital role in the generation of leaf shape. PCD also plays an essential role during reproductive development in plants. As a result of PCD, in many unisexual flowers, male and female organ growth is initiated and inappropriate organs are aborted during floral development. Moreover, the H_2O_2 hypersensitive response (HR) activates PCD, which is essential for limiting a plant pathogen's nutrient supply.

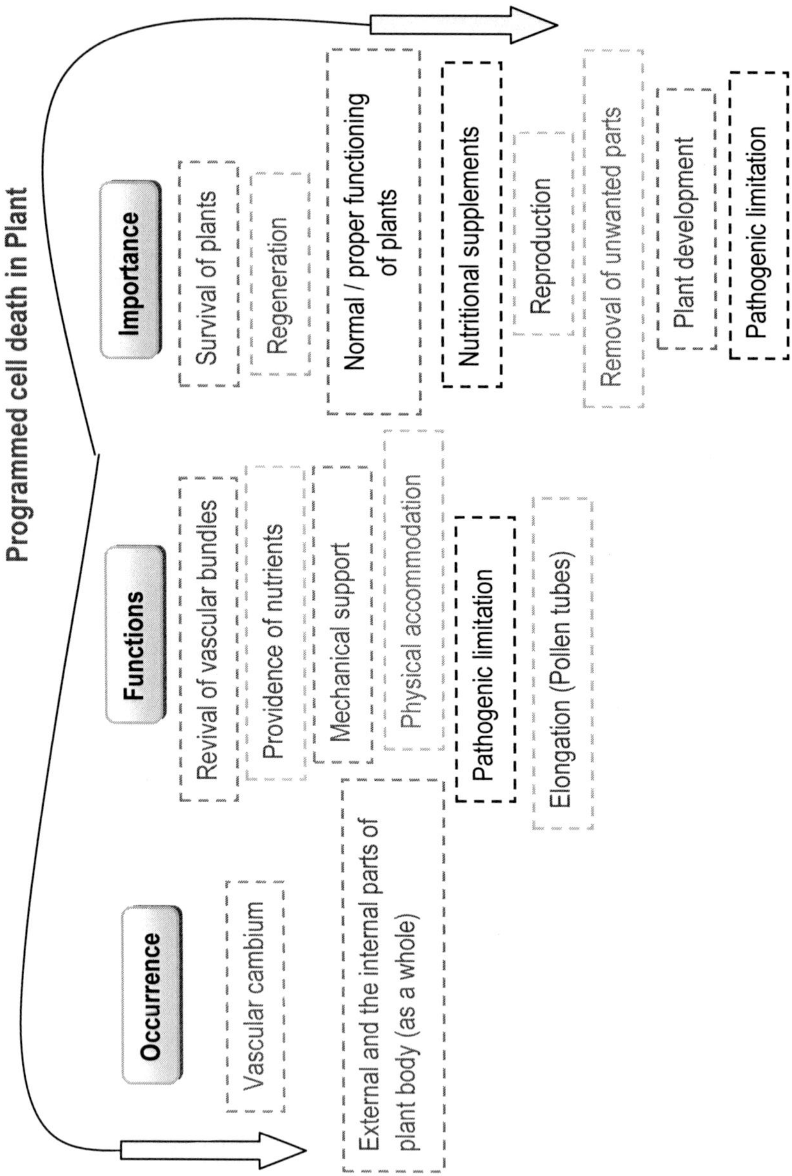

40. Outline of programmed cell death in plants.

Conclusions

The process of PCD is essential for ensuring the proper development of plants, as well as for providing a robust defense response against invading pathogens. In the quest to identify the signal for hypersensitive response (HR) induction, it has been determined that H_2O_2 is rapidly produced by plant cells in culture during the HR as part of a phenomenon referred to as oxidative burst; enhancing its production leads to a dramatic increase in PCD. Although the global mechanism for PCD control and execution used by plants remains unknown, it is intriguing that xylogenesis, toxin induced cell death, and some examples of the HR may occur via a mechanism that shares some similarities with apoptosis in animal cells. The execution of the steps used by the HR are similar to those of apoptosis and the question as to whether PCD arose independently in the plant kingdom or whether the existence of a common ancestor from which the process of PCD have been derived still needs to be determined. Overall, PCD directly or indirectly affects whole plants and ensures their survival.

References

Drew, MC, He, CJ and Morgan, PW. 2000. "Programmed cell death and aerenchyma formation in roots". *Trends in Plant Science*, No. 5: 123-127.

Dangl, JL, Dietrich, RA, Thomas, H. 2000. "Senescence and programmed cell death". In B Buchanan, W Gruissem, R Jones (eds.), *Biochemistry and Molecular Biology of Plants*. American Society of Plant Physiologists Press: Rockville, MD, pp 1044-1100

Vacca, RA, de Pinto, MC, Valenti, D, Passarella, S, Marra, E and De Gara, L. 2004. "Production of reactive oxygen species, alteration of cytosolic ascorbate peroxidase, and impairment of mitochondrial metabolism are early events in heat shock-induced programmed cell death in tobacco Bright-Yellow 2 cells". *Plant Physiology*, No. 134: 1100-1112.

Lam, E. 2004. "Controlled cell death, plant survival and development". *Nat. Rev. Mol. Cell Biol.*, No. 5: 305-315.

Montillet, JL, Chamnongpol, S, Rusterucci, C, Dat, J, Van de Cotte, B, Agnel, J-P, Battesti, C, Inze, D, Van Breusegem, F and Triantaphylides, C. 2005. "Fatty acid hydroperoxides and H_2O_2 in the execution of hypersensitive cell death in tobacco leaves". *Plant Physiology*, No. 138: 1516-1526.

Overmyer, K, Brosche, M, Pellinen, R, Kuittinen, T, Tuominen, H, Ahlfors, R, Keinanen, M, Saarma, M, Scheel, D and Kangasjarvi, J. 2005. "Ozone-induced programmed cell death in the Arabidopsis radical-induced cell death 1 mutant". *Plant Physiology*, No. 137: 1092-1104.

Danon, A, Miersch, O, Felix, G, op den Camp, RGL and Apel, K. 2005. "Concurrent activation of cell death: regulating signaling pathways by singlet oxygen in *Arabidopsis thaliana*". *Plant Journal*, No. 41: 68-80.

Frequently Used Acronyms

1-Napthaleneacetic acid	NAA
2,4-dichlorophenoxyacetic acid	2,4-D
6,4 pyrimidine-pyrimidone	6,4PP
ABA-responsive element	ABRE
Adenosine diphosphate	ADP
Alternative oxidases	AOX
Arabidopsis thaliana Stress Responsive Gene Database	ASRGD
Arsenic	As
Ascorbate peroxidase	APX
Ascorbic acid or ascorbate	AsA
Argininosuccinate	Asa
Adenosine triphosphate hydrolase	ATPase
ATP-binding cassette transporters	ABC transporter
Basic leucine zipper domain	bZIP
Bicarbonate ion	HCO_{3-}
C-repeat binding factor transcriptional activator	CBF/DREB1
Brassino steroids	BR
Bisphenol A	BPA
Ca^{2+} dependent protein kinases	CDPKs
Calcineurin B-like proteins	CBLs
Calcium	Ca^{2+}
Cation diffusion facilitator	CDF
Calmodulins	CaMs
Carotenoids	Car
Carbon : Nitrogen	C : N
Catalase	CAT
Centimeter	cm
Cold-regulated gene	COR
Comité International des Plastiques en Agriculture	CIPA
Cyclobutane pyrimidine dimers	CPDs
Cytochrome	Cyt
Dehydration responsive element-binding protein	DREB
Dehydroascorbate reductase	DHAR
Diacylglycerol	DAG
Double-strand DNA breaks	DSBs

Effector-triggered Immunity	ETI
Electron spin resonance spectroscopy	ESR
Electromagnetic frequency radiation	EMF_r
Electromagnetic radiation	EMR
Electron transport chain compartments	ETCs
Electroencephalograph	EEG
Epigenomic quantitative trait loci	eqtl
Ethylenevinyl acetate	EVA
Ethylene butyl acrylate	EBA
Extremely low frequency	ELF
Europium	Eu^{3+}
Fructose-1,6-bisphosphatase	FBPase
Fourier transform ion cyclotron resonance mass spectrometry	FI-ICR-MS
Fourier Transform Infrared Spectroscopy	FTIR
bisphosphate carboxylase large chain	cbbl
Gene for protein betaglucosidase	β-glu
Gene coding for chitinase enzyme	chi-A
Gene coding for nitrite reductase enzyme	nirS
Gene coding for nitrous oxide reductase enzyme	nosZ
Genome-wide association studies	GWASs
Geomagnetic field	GMF
Glutaredoxin	GRX
Glutathione	GSH
Oxidized glutathione	GSSG
Glutathione peroxidase	GPX
Glutathione reductase	GR
Glutathione S-transferase	GST
Glutathione synthase	GS
Global positioning system	GPS
Heat shock proteins	HSPs
Hydrogen peroxide	H_2O_2
Hydroxyl ions	OH^-
Hydrogen sulfide	H_2S
Hypersensitive response	HR
High Density Polyethylene	HDPE
Inositol-1,4,5-triphosphate	IP3
International Agency for Research on Cancer	IARC
Ionizing radiation	IR
International Telecommunications Union	ITU
Jasmonic acid	JA

Kilodalton	kDa
Kyoto Encyclopedia of Genes and Genomes	KEGG
Lanthanum	La^{3+}
Late embryogenesis abundant proteins	LEA
Magnetoencephalography	MEG
Magnetic Nanoparticles	MNPs
Mega Hertz	MHz
Malondialdehyde	MDA
Metallothioneins	Mts
Metabolomics Standards Initiative	MSI
Methylglyoxal	MG
Multidrug and toxin efflux protein	MATE
Master cell cycle regulation transcription factors	MYC
Mass defect filtering	MDF
Mitogen-activated protein kinase	MAPK
Mobile frequency	MF
Monodehydroascorbate reductase	MDHAR
My elob lastosis transcription factor	MYB
N-acetylcysteine	NAC
NADP dependent malic enzyme	NADP-ME
Natural resistance-associated macrophage proteins	Nramp
Nicotianamine	NA
Nano meter	Nm
Nanoparticles	NPs
Naphthaleneacetic acid	NPA
Nicotinamide adenine dinucleotide phosphate	NADP
Nitrate reductase	NR
Nitrogen, phosphorus, and potassium	NPK
Oxygen	O_2
PAMP-triggered immunity	PTI
Pattern recognition receptors	PRRs
Pathogen-associated molecular patterns	PAMPS
Peroxidase	POX
Peroxiredoxin	PRX
Peroxisomal membrane polypeptides	PMPs
Phenylalanine ammonia lyase	PAL
Phosphatidylglycerol	PG
Phosphoenolpyruvate	PEP
Photo System	PS
Photosystem II	PSII
Photosynthetically active radiation	PAR

Plant metabolic network	PMN
Pattern-triggered immunity	PTI
Plant Stress Protein Database	PSPDB
Polyphenol oxidase	PPO
Polyunsaturated fatty acids	PUFA
Phytochelatins	Pcs
Polyethylene	PE
Polyvinyl chloride	PVC
Polystyrene	PS
Polyacrylamide	PA
Polycarbonate	PC
Proteomic quantitative trait loci	pQTL
Programmed cell death	PCD
Radio frequency	RF
Radiofrequency electromagnetic fields	RF-EMF
Reactive oxygen species	ROS
Relative water content	RWC
Resistance	R
Ribulose-1,5-bisphosphate	RUBP
Ribulose-1,5-bisphosphate carboxylase/oxygenase	Rubisco
Sedoheptulose-1,7-bisphosphatase	SBPase
Single-strand DNA breaks	SSBs
Singlet oxygen	$^1O_2$
Specific absorption rate	SAR
Sodium nitroprusside	SNP
Stress Responsive Transcription Factor Database	SRTFDB
S-nitroso glutathione	GSNO
Soil-beta-glucosidase	S-β-GC
Soil-urease	S-UE
Superoxide anions	O^{2-}
Superoxide dismutase	SOD
Thioredoxin	TRX
Transcription factors	TFs
Ultra violet	UV
United States Environmental Protection Agency	USEPA
WRKY-DNA-binding proteins transcription factor	WRKY
Waterlogging	WL
Zinc–iron permease	ZIP
Zero magnetic field condition	ZMF

CONTRIBUTORS

GEETHA N, PhD
Department of Studies in Biotechnology, University of Mysore,
Karnataka, INDIA

S DUHAN, PhD
Department of Botany,
PG College
Chaudhary Ranbir Singh University
INDIA

S VIJAYAN, PhD
School of Biosciences, Mahatma Gandhi University
Kerala, INDIA

D JAIN, PhD
Department of Bioscience and Biotechnology, Banasthali Vidyapith,
Rajasthan, INDIA

G GARG, PhD
School of Biotechnology, Gautam Buddha University,
Uttar Pradeah, INDIA

R TRIPATHI, PhD
Department of Bioscience and Biotechnology, Banasthali Vidyapith,
Rajasthan, INDIA

S DUBEY, PhD
Amity Institute of Biotechnology,
Amity University, Uttar Pradesh, INDIA

MS JISHA, PhD
School of Biosciences, Mahatma Gandhi University
Kerala, INDIA

M JAIN, PhD
School of Biochemistry, Devi Ahilya Vishwavidyalaya,
Indore, INDIA

MANJU SHRI, PhD
Central Uiversity of Gujarat,
Gujarat, INDIA

M THAMPI
School of Biosciences, Mahatma Gandhi University
Kerala, INDIA

B JAMRA
School of Biochemistry, Devi Ahilya Vishwavidyalaya,
Indore, INDIA

ABHIJITH P
Department of Studies in Biotechnology, University of Mysore,
Mysuru, Karnataka, INDIA

S KATARIA, PhD
School of Biochemistry, Devi Ahilya Vishwavidyalaya,
Indore, INDIA

BHAVYA G, PhD
Department of Studies in Biotechnology, University of Mysore,
Mysuru, Karnataka, INDIA

P JANMEDA, PhD
Department of Bioscience and Biotechnology, Banasthali Vidyapith,
Rajasthan, INDIA

P CHAUDHARY
Department of Bioscience and Biotechnology, Banasthali Vidyapith,
Rajasthan, INDIA

S BHATI, PhD
School of Biotechnology, Gautam Buddha University,
Uttar Pradeah, INDIA

N D DHANRAJ,
School of Biosciences, Mahatma Gandhi University
Kerala, INDIA

P DUHAN, PhD
Department of Bio & Nano Technology,
Chaudhary Ranbir Singh University
INDIA

R S BAPNA, PhD
Laxmi Narain College of Technology,
Indore, INDIA

L DARNEI
Department of Botany, Haflong Government College,
Assam University, Haflong, INDIA

V Gupta
Indian Institute of Forestry Management,
Bhopal, INDIA

RK UPADHYAY, PhD
Department of Botany, Haflong Government College,
Assam University, Haflong, INDIA

INDEX

A

alpha-amylases, 235
Agriculture, 70, 84, 153, 155, 157, 203, 219, 223, 240
Aquaporins, 136, 150,151
Anatomical response, 138, 150
Anaerobic, 81, 133, 139, 141, 239
Adaptation, 2, 22, 41, 80, 91, 138, 220
Alkaloids, 4, 178, 192
Aliphatic amino acids, 4
Arbuscular mycorrhizal fungi, 7
Antioxidant, 14, 37, 40, 47, 49, 66, 68, 73, 86, 113, 117, 141, 178
Anthocyanin, 14, 16, 176, 177
Artemisinin, 18
Artemisia, 18, 16
Anti-inflammatory, 8-10
Anti-cancer, 10
Abscisic acid, 64, 69, 73, 80, 82, 87, 113, 221
Auxin, 48, 105, 107, 113, 138

B

Beta-amylases, 235
Bioactive constituents, 1, 6, 11
Biochemical pathways, 1, 192, 198
Biogenesis, 1, 58
Biotic stresses, 172, 197
Bug, 4
Biochemical changes, 63, 73, 183, 229, 233
Biocatalyst, 6
Biomolecules, 12, 13, 40, 79, 234
Bacoside, 17
Betulinic acid, 18, 19
Betalain, 17

C

Cosmetic, 1, 156
Catabolism, 7, 142
Competitive pathway, 7
Combined stress, 172, 174, 181, 184
Cell death, 234, 239, 241, 245
Callus culture, 6
Cell culture, 4, 6, 22
Carvacrol, 8
Chemotherapeutic, 9
Cirrhosis, 8
Curcumin, 9, 11
Cinchona tree, 9
Cytotoxic activity, 10, 108, 115, 179, 220
Contamination, 12, 105, 161, 163
Caffeic acid, 10
Cold, 15, 62, 83, 157, 191, 218, 239
Chilling, 40, 42, 54, 81, 171
Chlorophyll, 18, 40, 46, 63, 77, 105, 143, 221
Crop management, 192

D

Diosgenin, 17
Drought, 12, 17, 19, 35, 38, 40, 63, 68, 171, 173, 175
Dehydrate, 20

E

Environment, 191, 198, 218, 229, 240
Electromagnetic radiations, 209, 211, 229, 233, 236
Expression, 2, 11, 42, 47, 68, 107, 109, 142, 183
Ecosystem, 22, 183
Exposure, 37, 49, 71, 106, 109, 136, 176

Emission, 211, 117
Edaphic, 4, 12

F
Flood- tolerant genes, 142
Flux, 7, 67, 218
Fingerprinting, 193, 195, 202
Fisetin, 9
Fertilizers, 13, 104, 158, 163, 208
Flavonoids, 14, 18, 21, 38, 173, 177
Free radicals, 178, 220, 235
Fungi, 11, 62, 76, 114, 191

G
Glycolysis, 67, 133
Globe, 1, 11
Genes, 11, 35, 45, 63, 106, 111, 163
Genetic stability, 7
Geotropism, 7
Growth, 7, 11, 35, 45, 84, 105, 153
Glycyrrhizin, 8
Ginsenosides, 14, 16
Glucosinolates, 19
Geographical, 22

H
Herbs, 4
Heat stress, 77-80
Hatchling, 4
Hairy root, 6, 7
Hormone, 7, 47, 70, 72, 73, 108, 181
Host, 2, 7, 11, 22, 203
Heavy metals, 16, 45, 47, 63, 109
Hypericin, 16, 18, 19
Homeostasis, 11, 38, 41, 42, 82, 114, 192, 239
Hyperaccumulator, 109, 115
Hypersensitive response, 234, 245

I
Immobilization, 6, 154, 164
Inoculation, 7
Infection, 7, 192, 195-197, 207, 223, 240, 243
Ingenol, 9
Insecticides, 9, 38
Irradiation, 14, 15, 174, 178, 220
Indole, 20, 22, 108, 113
Immune system, 12, 52, 197

J
Jasmonates, 2, 16, 37, 111

L
Linalol, 8
Lipid peroxidation, 41, 47, 63, 68, 74, 77, 85, 182, 235
Lepidine, 17

M
Metabolites, 1-11, 14, 16, 176, 177, 191-198, 200-203
Metabolomic Tools, 192
Membrane stability, 79, 142
Membrane injury, 63, 181
Matrix, 6, 39, 40
Meristem, 6, 107
Metabolic engineering, 4, 7, 22
Mucilage, 8
Marshmallow, 8
Melatonin, 16
Membranes, 34, 35, 37-39, 43, 57, 66, 73, 74, 77
Menthol, 8
Menthone, 8
Microwave radiation, 230, 234
Mulching, 153, 156-158, 165, 166
Mutations, 48, 202, 230

N
Non-ionizing radiation, 230
Nourishment, 4
Nicotine, 10

O
Oxidative stress, 34, 37, 38, 40, 41, 44, 47

P
Photosynthetic rate, 78, 108, 136, 182

Pathogenic, 2, 7, 52, 196, 240
Pharmaceuticals, 2, 4
Polyamines, 2, 15, 74, 80, 178, 179, 181, 197
Phenolics, 4, 18, 73, 74
Phytochemicals, 4, 6, 192
Parasites, 11, 192
Pesticides, 1, 11, 38, 45, 70, 104, 172, 223
Pollutants, 11, 12, 38, 45, 70, 160, 172
Plant hormones, 73, 82, 113, 173, 179, 181
Plant pathogen interaction, 195
Photosynthetic pigments, 16, 77, 176
Phytoalexin, 17, 196, 197
Phytochelatins, 104, 113, 114
Photosynthesis, 15, 20, 38, 42, 105, 134, 136, 143, 176, 181, 220
Phytoremediation, 105
Potassium, 18, 163
Programmed cell death, 239, 240, 243
Pseudohypericin, 18

Q
Quassinoids, 8
Quinine, 9
Quercetin, 18

R
Radiation, 11, 14, 40, 48, 51, 79, 80
Rhabdomyosarcoma, 9
Reserpine, 9, 20
Root nodules, 140
Resveratrol, 17
Respiration, 20, 38, 66, 79, 107, 143
Reactive oxygen species, 12, 73, 87, 106, 116, 137, 178, 220

S
Secondary metabolites, 1, 4, 6, 11, 16, 22, 171, 196
Stimuli, 11, 37, 120, 219
Salicylic acid, 2, 73, 82, 113
Shikimic acid, 4
Senescense, 64, 66, 82, 105, 137
Signal transduction, 71, 104, 106, 192
Signaling molecule, 37, 54, 72, 82, 111, 191, 196
Salvianolic acid, 19
Saikosaponins, 19
Solasodine, 20
Stomata, 12, 35, 41, 47, 66, 77, 171
Salinity, 1, 12, 17, 19, 22, 40, 63, 84, 172, 220

T
Temperature, 11, 13, 15, 19, 35, 67, 80, 158, 214
Therapeutics, 1, 31
Terpenoids, 4, 14
Tricarboxylic acid cycle, 4
Tissue culture, 26, 30
Transduction, 35, 37, 51, 106, 111, 241
Thymol, 8
Tachycardia, 9
Thyrotoxicosis, 9
Tomato plant, 10, 72, 83, 136,197
Textile, 11
Taxol, 15, 17
Temperature, 1, 11-19, 35, 158, 166, 172, 198, 214, 215, 220, 239
Transpiration rate, 67, 136, 137

U
Ultra violet, 38, 48, 51, 56, 129
UV-B radiation, 40, 48, 49, 51, 54, 171-179, 181-183

V
Vincristine, 9, 14, 20
Vinblastine, 8, 14
Vanillin, 11
Variation, 1, 11, 15, 20, 48, 106, 193, 198, 199, 233

W
Wavelength, 14, 229, 230
Waterlogging, 62, 63-67, 86, 87, 133, 134, 139, 140, 220
Water deficit, 68, 69,97, 174, 176, 178, 182, 183

X
Xanthotoxin, 17

Y
Yeast extract, 6
Yield, 45, 49, 62-67, 70, 77

Z
Zingiberene, 14